Compressor Instability with Integral Methods

Y.K. Eddie Ng, Ningyu Liu

Compressor Instability with Integral Methods

With 67 Figures and 11 Tables

Springer

Dr. Y.K. Eddie Ng
Nanyang Technological University
School of Mechanical &
Aerospace Engineering
College of Engineering
50 Nanyang Avenue
Singapore 639798
Singapore

Dr. Ningyu Liu
National University of Singapore
Department of Mechanical Engineering
10 Kent Ridge Crescent
Singapore 119260
Singapore

ISBN   978-3-642-09147-6   e-ISBN   978-3-540-72412-4

Springer is a part of Springer Science+Business Media
springer.com
© Springer-Verlag Berlin Heidelberg 2010

Cover design: WMX Design, Heidelberg

Preface

The CFD were significant tools for the engineers and analysts in the turbomachinery industries. The development of computational resources and computational methods enables us perform these CFD calculations quickly, and reliance on less component testing.

By the demand-pull of turbomachinery industries, the most comprehensive flow simulation models and improved methods have been established to achieve the ever-more flow feature and geometrical complexity and fidelity. Recently, several powerful and easy-to-use CFD software tools have been developed and dominate throughout the fluid machinery industry.

Because of the use of CFD is depended on computational sources, the experience of the software users, the options of numerical methods, physical modes, geometry definition, etc, the demand to use a simplify tool to predict and analyze integrated into the turbomachinery components encourages us to develop such a single set of formulations.

The main topic and scope of this book is to develop some simplify tools to predict and analyze performance of the axial compressor with inlet distortion. The integral method is the typical one in these simplify tools. In this book, the integral method will be introduced with a detail derivation, application and development. The integral method is firstly proposed by Kim et al. at 1996 aimed to describe the problems of distorted inlet flow propagation in axial compressors. In this method, the blade rows in multistage axial compressor are replaced by their equivalent force field, and the Navier-Stokes equation was simplified by using integral technique. Despite the great simplifications adopted in the integral method, the multistage axial compressor with large inlet distortion, including back flow can be analyzed and a qualitative trend of distortion propagation can be described successfully. The integral method provides the useful information about the performance of the axial compressor with inlet distortion, which is meaningful to engineering application in the design and analysis of turbomachinery.

Another simplify tool to predict the performance of axial compressor is Greitzer's B-parameter method, which will be introduced in Chap. 5. The B-Parameter method is a simple and useful tool to apply in analyzing the stall and surge characteristics for a compressor system as a whole. The detail analysis and cases studies are also provided.

To ease the readers in using the methods mentioned in this work, some detail derivation and several Fortran source programs are included. Readers can test the examples provided in each chapter and/or, further, solving some other relevant problems.

The Taguchi parameter study cases in Chap. 3 and Chap. 5 equip readers through the analysis procedure step-by-step to reach the conclusion. One can therefore easily design a similar example to practice it and then know it well.

Contents

Reviews

"This book is intended to be a reference material mainly for aeronautical engineering senior students, graduate-level students and practitioners in the aerospace industry with the essential background of college-level physics and integral calculus.

With the level of mathematics and aerodynamics involved in this book, this practical book, with its corresponding FORTRAN source codes to test associate examples and problem sets (both a novice and ranging from basic to complex), is ideal for upper-level undergraduate and graduate students in engineering to provide an opportunity for essential practice. It focuses on the improved integral methods applied to the distorted flow analysis. Throughout the text, numerous in-flight examples from both the commercial and military compressors show readers how the concepts and effective calculations are applied to real-life operations. The unique book also serves as a reference for design engineers who want a set of simple and fast methods to predict the performance of axial compressor."

Prof. Sadanari Mochizuki, Ph.D.
Editor-in-Chief, The International Journal of Rotating Machinery
Department of Mechanical Systems Engineering, College of Engineering,
Tokyo University of Agriculture and Technology, 2-24-16 Nakacho, Koganei,
Tokyo 184-8588, Japan

"This book presents a mathematical treatment of axial compressor response to inlet distortion. Computational results show the requirements for stable operation and the consequences of unstable operation in terms of growth of the distorted region through the compressor, the reduction of mass flow and the change in compressor performance. The analysis shows that the compressor response depends on the magnitude of the inlet distortion, the inlet flow angle and the ratio of lift/drag coefficients of the blade. This analysis helps to understand the details of downstream response to inlet distortion in terms of mass flow reduction and changes in the level of distortion through the blade-rows of a multistage axial compressor.

The mathematical approach uses the continuity equation and the equations of motion in two dimensions. These equations are expressed in terms of differential distortion parameters which are integrated and substituted into the differential equations of motion. Numerical integration is performed by the fourth order Runge-Kutta technique, and solutions are presented for various levels of distortion. In Chap. 2, results of sample calculations shown in Figs. 2.4, 2.5 and 2.6 illustrate the capability of the analysis toward providing a good understanding of the details and causes of instability. Figures 2.7 and 2.8 are good illustrations of the ability of the method to replicate the compressor characteristics in stall. This is very good.

The book could be a valuable reference toward understanding axial compressor response to inlet distortion through its mathematical and analytical treatments."

Professor Dr. Ronald C. Pampreen, Ph.D.
Professional Development Programme, Concepts ETI, Inc.
Concepts NREC, Massachussetts, USA

"This book manages to cover an analysis for the effect of distorted inlet flow propagation on the rotating stall and surge in axial compressors in a relatively compact way and could form a useful introductory text. It provides an insight into the basic phenomena controlling the compressor flow instabilities, and to reveal the influence of inlet parameters on rotating stall and surge. These useful technical details are interesting. The authors frequently illustrate the formulae they provide with specific numerical examples which helps to put them in context. In all, this is a book written with much time, effort and consideration. It is a one of a kind advance engineer book written by Singaporean for the local engineers in mind."

Dr. Chen Chuck, Ph.D.
Principal Scientist,
The Boeing Company,
Seattle, WA, USA

"This book gives basic guidance on a much broader range of compressor stall and surge topics than is usually seen in books aimed at non-engineers. It can be used as a specialized topic of senior undergraduate or graduate study in the flow instabilities analysis, design and testing the influence of inlet parameters of axial compressor. The corresponding FORTRAN source codes to test associate examples and problem sets (both a novice and ranging from basic to complex), provides additional good material and is ideal for graduate engineering students to allow an opportunity for essential practice"

Prof. Dr. J.C. Misra, Ph.D., D.Sc.
President, Mathematical Sciences, Indian Science Congress
Editor-in-Chief, Mathematical Sciences Series
Former Head, Department of Mathematics
Former Head, School of Medical Science & Technology
Centre for Theoretical Studies, Indian Institute of Technology, Kharagpur, India

"I particularly like the way the 5 chapters (albeit a concise book) and all the subheadings have been clearly listed in the Contents so that the reader should have little trouble in discovering where to look for the topic of interest. The section on 'in-flight example for fighter jet compressor' is one of the better written sections as it provided practical points in showing readers how the concepts and effective computations are applied to real-life operations' ... The authors are to be warmly congratulated for having written a very useful book. I strongly recommend this textbook to professionals in industrial power and propulsion groups."

Emeritus Professor Wen-Jei Yang, Ph.D., P.E.
Department of Mechanical Engineering and of Biomedical Engineering
University of Michigan, Michigan, USA
Fellow, ASME
Founding President: *The Pacific Center of Thermal-Fluids Engineering*

"Based on Kim and Marble's integral method, this book further develops this method through the proposing and applying a critical distortion line. This line is applied successfully on the stall prediction of in-flight compressor due to flamming of refueling leakage near inlet. It provides a typical real and interesting example of compressor stall and surge operation. An excellent, concise and practical guide for engineers in all the rotating machinery professions."

Professor WG Park, Ph.D.
School of Mechanical Engineering,
Pusan National University, Korea

"This book provides an insight into the basic phenomena controlling the compressor flow instabilities, revealing the influence of inlet parameters on rotating stall and surge. Based on its purpose, each chapter is constructed. In other words, an extended Greitzer's instability flow model, the well-known B-parameter model applied for analyzing the stall and surge characteristics parametrically using Taguchi method are included. Thus researchers and gradient students can get much information through the book."

Professor Shuichi TORII, Ph.D.
Department of Mechanical System Engineering,
Kumamoto University, Japan

"This book packs a big punch for its size. Each chapter is written concisely and ends with a list of sources of additional information for those who are interested where appropriate. The book is understandable and useful to the engineering community."

Dr. Akira GOTO, Ph.D.
Director and Board Member, Ebara Research Co., Ltd
Deputy Division Executive,
Fluid Machinery Development Division
Fluid Machinery and Systems Company
Ebara Corporation, Tokyo, Japan

"The book is well worth the price and should be useful reading for rotating machinery professionals ... This book's title accurately reflects its contents and it has excellent chapters as to make it very worthwhile reading."

Dr. J.T. Kshirsagar, Ph.D.
Head of R & D Technology, Kirloskar Brothers Limited, Pune, India
Distinguished Visiting Professor at I.I.T. Madras, I.I.T. Roorkee,
I.I.T. Bombay, India

"Flow instability, stall and surge are inter-related, and hence together constitute an intriguing phenomenon that governs axial compressor performance. This fascinating and wonderful book is unique in addressing the inlet flow distortion features and their unsteady propagation characteristics of compression process. The flow governing formulations of the various chapters are not only sufficiently rigorous to be instructive for aeronautics courses, but are also practically oriented."

Prof. Dr. Mohd. Zamri, Yusoff, Ph.D.
Dean, College of Engineering,
Universiti Tenaga Nasional, Malaysia

"This is an outstanding and interesting book ... cuts through the mystery and complexity of a subject that haunts many compressor designers and allows new integral approach to the flow instability prediction. It is a must read for both engineers and turbomachinery professions."

Dr. K.C. Ng, Ph.D.
Department of Research & Applications,
O.Y.L.R & D Center, Selangor, Malaysia

"This useable book is intended for advanced undergraduate and graduate students in mechanical and aerospace engineering. Among the features of the book are an emphasis on the role of quick integral method in assessing axial compressor performance; a timely review of flow instabilities; revisiting the famous Greitzer's B-parameter model as it applies to actuator-disc for the stall and surge flow characteristics; and highlighting the importance of improved Kim and Marble's integral method with a novel critical distortion line through a turbomachinery component in carrying out real computational tasks. It will prove to be a valuable resource to all who have an interest in the subject of fluid machines, and in particular, I recommend it highly to those of us who are not only aerodynamicists per se, but also need to be increasingly well informed on turbomachinery aerodynamic matters."

Professor Chyi-Yeou Soong, Ph.D.
Department of Aerospace and Systems Engineering,
Feng Chia University, Taiwan, ROC.
AIAA Associate Fellow,
Editor, Journal of Aeronautics, Astronautics & Aviation

I Foreword

I.1 About this Book

This book focuses on the numerical analysis for the effect of distorted inlet flow propagation on the rotating stall and surge in axial compressors. The purpose is to gain insight into the basic phenomena controlling these flow instabilities, and to reveal the influence of inlet parameters on rotating stall and surge.

The book starts from the confirmation and application of Kim's integral method [3], and then follows by a development to this method through the proposing and applying a critical distortion line (Chapter 1). This line is applied successfully on the stall prediction of in-flight compressor due to flamming of refueling leakage near inlet, a typical real and interesting example of compressor stall and surge operation (Chapter 2). Further, after a parametric study on the integral method and the distorted flow field of compressor using Taguchi method ([4] and [5]) (Chapter 3), a novel integral method is formulated using more appropriate and practical airfoil characteristics, with a less assumptions needed for the derivation of integral equations (Chapter 4).

Finally, in Chap. 5, as an extended work, the famous Greitzer's instability flow model ([1] and [2]), the well-known B-parameter model applied for analyzing the stall and surge characteristics, is studied parametrically using Taguchi method as an example to illustrate the stall and surge analysis through the whole compressor system.

I.2 Methods Used in this Book

In this book, a number of numerical methods are adopted. The main methods for numerical simulation and analysis are integral method and Greitzer's B-parameter model ([1] and [2]). The Taguchi quality control method ([4] and [5]) is used in parametric study, and the fourth-order Runge-Kutta method is applied in solving the partial differential equations. Otherwise, the Chebyshev curve fitting is used to formulize the previous experimental and numerical data. The commercial software is used here to obtain database for the purpose of comparison.

(i) The software used in this book is FASTFLO, which is introduced briefly below.

(ii) The Fortran program of integral method can be found in the Appendix of Chapter one.

(iii) The Fortran program of Chebyshev curve fitting can be found in the Appendix of Chapter four.

(iv) The Fortran program of Greitzer's B-parameter model can be found in the Appendix of Chapter five.

I.3 Research Application

(i) First of all, based on the study of integral method, a critical distortion line and corresponding distortion propagation factor are proposed to express the effect of the two main inlet parameters: the angle of flow and the distorted inlet velocity, on the propagation of distortion.

(ii) The practical field applications, such as the inlet conditions of flaming of leakage fuel during mid-air refueling process, are implemented to show the details of the integral method and critical distortion line used in analysis of the axial flow compressor behavior and the propagation of inlet distortion.

(iii) The results of parametric study by using Taguchi's quality control method indicate that the influence of major parameters on the inlet distortion propagation can be ranked as, the most one of the ratio of drag-to-lift coefficient, then the inlet distorted velocity coefficient, and the least one of inlet flow angle.

(iv) The study indicates that the previous integral method underestimated greatly the inlet distortion propagation. By using the newly developed integral method, an investigation is proceeded to present the effects of inlet parameters upon the downstream flow features with inlet distortion, including the inlet distortion propagation, the compressor critical performance and critical characteristic.

(v) Based on the parametric study of Greitzer's B-parameter model applied for analyzing the stall and surge characteristics, a total of four parameters B, G, K and L_C in the model are highlighted in order to establish the influence of each parameter on the whole compressor system. It is found that parameter K is the deciding factor in the position of the cross point whereas parameter B is the most important one in changing the length of time needed for the compressor to reach its steady state.

I.4 Acknowledgments

The work presented here was initiated with NTU-DSTA interests and collaborated with the Republic of Singapore Air Force (RSAF). The first author wants to acknowledge Prof. Frank Marble of California Institute of Technology, for bringing the problem to the author's attention and for his helpful discussion.

This book is dedicated to Mr. Lim Hong Ngiap, Dy. Director of Air Logistic Department, RSAF. His unwavering supports and completes faith in our dream during the last 15 years that this effort has taken is what that made it possible.

Finally, the authors would like to express their deepest appreciation to the research students, Ms. Siak Jia Huey and Ms. Tan Shin Yi for their great effort in the application of Taguchi method upon the parametric analysis of compressor, and the propulsion engineers of RSAF, Mr. Tan D.H.Y and Mr. Tan T.L. for their keen interests and helpful discussions.

I.5 Briefly Introduction of Software: *FASTFLO*

In this book, the software *Fastflo* Version 3.0 is adopted to obtain database as the direction and comparison throughout our numerical simulation.

Fastflo is a general-purpose package for the numerical solution of 2D or 3D partial differential equations (PDEs) by the finite element method. As a general PDE solver, *Fastflo*'s main advantage is its flexibility in specifying models and algorithms to solve them. The PDEs can be well known (such as fluid flow, solid mechanics, electromagnetism, eigenvalue problems) or non-standard equations encountered in scientific or industrial applications. Users are free to specify what equation to solve, to design the algorithm used for the solution, and to control the computations intelligently. *Fastflo* has its own mesh generator and post-processing capability.

Users present their PDE problem to *Fastflo* via two files, one for the mesh and one for the problem specification. Because *Fastflo* uses unstructured meshes, problems can be solved in complex geometrical shapes. If the PDE is time dependent, the user may develop an algorithm for time-stepping. If the PDE is nonlinear, then an appropriate iterative strategy needs to be implemented.

After any necessary time-stepping and iterative algorithms have been implemented, the user will have a set of linear PDEs to be solved at each time step or iteration. In *Fastflo*, these PDEs can have partial derivatives up to second order and vector or tensor coefficients.

Dependence of coefficients on spatial and other variables is managed by a mathematical expression capability with a wide range of operators. A global vector stack is used to store and manipulate the field variables.

More information about the software, please refer to the on-line Reference Manual or Tutorial Guide:

http://www.nag.co.uk/simulation/Fastflo/Documents/Tutorial/tutorial.htm

I.6 References

[1]   Greitzer E.M., 1980, Review: axial compressor stall phenomena. *ASME Journal of Fluids Engineering*, **102**: 134-151.

[2]   Greitzer, E.M. and Griswold, H.R., 1976, Compressor-Diffuser Interaction With Circumferential Flow Distortion, *Journal of Mechanical Engineering Science*, **18(1)**: 25-38.

[3]   Kim J.H, Marble F.E, and Kim C-J., 1996, Distorted inlet flow propagation in axial compressors. In *Proceedings of the 6th International Symposium on Transport Phenomena and Dynamics of Rotating Machinery*, Hawaii, **2**: 123-130.

[4]   Taguchi, G., 1993, Taguchi On Robust Technology Development: Bringing Quality Engineering Upstream, New York: ASME Press.

[5]   Taguchi, G., 1986, Introduction To Quality Engineering: Designing Quality Into Products And Processes, Japan: Asian Productivity Organization.

I.7 List of Publications Related to this Book

1)   Ng E.Y-K., Liu N., Lim H.N. and Tan T.L., 2005,A Improved Integral Method for Prediction of Distorted Inlet Flow Propagation in Axial Compressor, ***International Journal of Rotating Machinery***, **2**: 117-127.

2)   Ng E.Y-K., Liu N. and Tan S.Y., 2005, Numerical Parametric Study of Inlet Distortion Propagation in Compressor Using Integral Approach with Taguchi Method, ***International Journal of Computational Methods in Engineering Science and Mechanics***, **6**: 169-177.

3)   **Ng**, E.Y-K., Liu, N., Lim H.N. and Tan H.Y., 2005, A Sensitivity Study of the Distorted Inlet Flow in Axial Turbomachinery with Novel Integral Scheme, ***Journal of Computational Fluids Engineering*** (selected paper # K2 from ACFD5), **10(1)**: 51-55.

4)   Ng E.Y-K., Liu N. and Tan S.Y., 2004, Parametric Study of Greitzer's Instability Flow Model Through Compressor System Using the Taguchi Method, ***International Journal of Rotating Machinery***, Taylor & Francis, U.K., **10: 91-97**.

5)   Liu N., Ng E.Y-K., Lim H.N. and Tan T.L., 2003, Stall Prediction of In-Flight Compressor Due To Flamming of Refueling Leakage near Inlet, ***Computational Mechanics***, **30(5-6)**: 479 – 486.

6)   Ng E.Y-K., Liu N. and Tan S.Y., Study of Greitzer's B-Parameter Model Using ANOVA & Taguchi Method, *The ACFD5 Conference*, Busan, Korea, October 2003.

7)   Ng E.Y-K., Liu N. and Siak J.H., Parametric study of inlet distortion propagation in compressor with integral approach and Taguchi method, *The 7th, ASIAN Int. Conference on Fluids Machinery (AICFM)*, Fukuoka, Japan, Oct. 2003.

8)   Liu N., Ng E.Y-K., Lim H.N. and Tan T.L., Prediction of inlet distortion induced axial compressor flow field instability in mid-air refueling, Accepted by *The 2003 Joint ASME-JSME Fluids Engineering Summer Conference (FEDSM2003–45393)*, Hawaii, July 2003.

9)   Ng E.Y-K., Liu N., Lim H.N. and Tan T.L., 2002, Study on The Distorted Inlet Flow Propagation In Axial Compressor Using An Integral Method, ***Computational Mechanics***, 30(1): 1-11.

10)   Liu N., Ng E.Y-K., Lim H.N. and Tan T.L., An applicability study of integral method in axial compressor with a distorted inlet flow during mid-air refueling process, In *Proceedings of RSAF ATS seminar*, Singapore, September 2002.

11)   Ng E.Y-K., Liu N., Lim H.N. and Tan T.L., Computation of axial flow compressor with intake flow distortion, In *Proceedings of The 2002 Joint US ASME-European Fluids Engineering Summer Conference*, FEDSM2002-31202, Montreal, Canada, July 2002.

I.8 Nomenclature

Symbols:

a speed of sound

A area of the compressor duct

B dimensionless parameter, $B = \dfrac{U}{2a}\sqrt{\dfrac{V_P}{A_C L_C}}$

C compressor pressure rise

C_x compressor axial velocity

C_{SS} Steady-state compressor pressure rise

C_l lift coefficient

C_d drag coefficient

F throttle pressure drop

$F_{x,0}$ x- direction force in the undistorted region

$F_{y,0}$ y- direction force in the undistorted region

F_x x- direction force in the distorted region

F_y y- direction force in the distorted region

$F_\perp$ force component normal to the local velocity vector

$F_{//}$ force component parallel to the local velocity vector on rotor blade

G dimensionless parameter, $G = \dfrac{L_T A_C}{L_C A_T}$

k polytropic exponent

K dimensionless parameter, $K = \dfrac{A_C^{\,2}}{A_T^{\,2}}$

K_0 constant, $K_0 = K_1 + [1 - K_1/\alpha]\alpha_0$

K_1 constant, $K_1 = \dfrac{\delta\alpha}{\pi R}$

K_2 velocity parameter, $K_2 = \dfrac{K_1(K_1 - K_0)}{(\alpha - K_1)^2}$

K_3 velocity parameter, $K_3 = 1 + \dfrac{K_2(K_1 - K_0)}{\alpha - K_1}$

k_D drag coefficient
k_L lift coefficient
k_D/k_L the ratio of drag-to-lift
L effective length of equivalent duct
$\dot{m}$ mass flow rate
M compressor axial flow coefficient, C_x/U
n total stage number of compressor
p static pressure

P_{02}/P_{01} ratio of total pressure at outlet to inlet

P_1 inlet non-dimensional static pressure

P_2 non-dimensional static pressure

ΔP non-dimensional pressure difference in the range of [0, x],

$$\Delta P = \frac{p(x) - p(0)}{\rho U^2}$$

R compressor rotor mean radius
Re Reynolds number
t time
U mean rotor velocity
U_0 x- component of referenced inlet velocity
V volume
V_0 y- component of referenced inlet velocity
u x- component of non-dimensional distorted velocity
v y- component of non-dimensional distorted velocity
u_0 x- component of non-dimensional undistorted velocity
v_0 y- component of non-dimensional undistorted velocity
w_1 non-dimensional undistorted velocity at inlet
w_2 non-dimensional undistorted velocity at outlet
x axial coordinate

y circumferential coordinate

$Y(x)$ function expression of a symmetric line of distorted region

α distorted velocity coefficient in x-direction

α_0 undistorted velocity coefficient in x-direction

$\bar{\alpha}$ wing section angle of attack

β distorted velocity coefficient in y-direction

β_0 undistorted velocity coefficient in y-direction

$\bar{\gamma}$ tangent of flow angle in inlet, $\bar{\gamma} = tan\,\theta_0 = V_0/U_0$

γ ratio of specific heats

$\Delta\xi$ increment of distorted region from inlet to outlet, $\Delta\xi = \xi(n) - \xi(0)$

δ circumferential extension of distorted flow

η coordinate of transformed coordinates system, $\eta = \dfrac{y - Y(x)}{\delta}$

θ local flow direction

θ_0 flow angle at inlet

$\bar{\lambda}$ length of a single stage in x-direction

λ length of blade in x-direction

μ critical distortion factor

$\xi(x),\ \xi$ relative circumferential extension of distorted flow,

$$\xi(x) = \frac{\delta(x)}{\pi R} = K_1/\alpha(x)$$

$\Delta\xi$ size increment of distorted region from inlet to exit, $\Delta\xi = \xi(L) - \xi(0)$

ρ density

ψ local flow direction relative to a rotor, $tan\,\theta = (\omega R - v)/u$

$\theta^*,\ \psi^*$ referential local flow directions relative to a stator and a rotor

ϕ The non-dimensional velocity rise

ψ_p The non-dimensional pressure rise

φ distortion propagation factor

σ ratio of rotor blade speed to y- component of referenced inlet velocity $\sigma = \dfrac{\omega R}{V_0}$

ω rotor angle velocity

ωR rotor blade speed based on mean radius

Γ distortion level, $\Gamma = 1 - \dfrac{\alpha}{\alpha_0}$

ε relative change rate of given function f in the range of $[0, x]$,

$$\varepsilon(f) = \frac{f(x) - f(0)}{f(0)}\%$$

Φ *non-dimensional mass flow rate,* $\Phi = \dfrac{(\alpha - \alpha_0)\xi + \alpha_0}{\gamma\sigma}$

τ compressor flow field time constant

Superscripts:
r *rotor*
s *stator*
$\sim$ nondimensionalized variable

Subscripts:
O initial nondimensionalized value at surge point (or stall point)
C compressor
P plenum
T throttle
x x-axial direction

Chapter 1
Study on the Propagation of Inlet Flow Distortion in Axial Compressor Using an Integral Method

An integral method is presented in this chapter. The integral method is proposed by Kim et al. at 1996 aimed to describe the problems of distorted inlet flow propagation in axial compressors. In this method, the detained blade rows are replaced by their equivalent force field, and the Navier-Stokes equation can be simplified by using integral technique. Despite the great simplifications adopted in the integral method, the multistage axial compressor with large inlet distortion, including back flow can be analyzed and a qualitative trend of distortion propagation can be described successfully. The integral method provides the useful information about the performance of the axial compressor with inlet distortion, which is meaningful to engineering application in the design and analysis of turbomachinery.

In this chapter, by applying the integral method, the effects of the parameters of inlet distortions on the trend of downstream flow feature in compressor calculated numerically is described. Other than the ratio of drag-to-lift coefficients of the blade and the angle of flow, the value of distorted inlet velocity is found to be another essential parameter to control the distortion propagation. Based on this calculation, a critical distortion line and corresponding distortion propagation factor are proposed to express the effect of the two main inlet parameters: the angle of flow and the distorted inlet velocity, on the propagation of distortion. From the viewpoint of compressor efficiency, the distortion propagation is further described by a compressor critical performance. The results present a useful physical insight of compressor axial behavior and asymptotic behavior of the propagation of inlet distortion, and confirm the active role of compressor in determining the velocity distribution when compressor responds to an inlet flow distortion.

1.1 Introduction

In the operation of aircraft jet-engine, it is important to understand the aerodynamic response resulting from an inlet flow distortion. An inlet flow distortion may cause the rotating stall or even surge, or a combination of them. If the blades fail to produce required loading, catastrophic damage to the complete engine would result. To avoid stall and surge of compressor due to flow distortion, and understand further, the inlet distortion and its propagation effects have received much attention over the years. In 1955, Harry and Lubick [7] investigated the turbojet engine in an altitude chamber at the NACA Lewis laboratory to determine the effects of a wide range of uneven inlet-air pressure distributions on transient characteristics and stall phenomenon.

The flow non-uniformities due to inlet distortion are commonly grouped into radially varying steady state, circumferentially varying steady state and unsteady distortions [5]. In many situations, the principal loss in stall margin can be regarded as due to one of the above three groups. Among them, the flow with circumferential distortion introduces new phenomenon into the fluid dynamic analysis of the compressor behavior and has gained much attention by many researchers. One of the methods used is a linearized approach, which provides quantitative information about the performance of the compressor in a circumferentially non-uniform flow. Several models, such as the parallel compressor model and its extensions, are used to assess the compressor stability with inlet distortion. The numerical solution of the time dependent nonlinear inviscid equations of motion is another method to study the problem of compressor stability in a distorted flow.

In 1980, Stenning [11] presented some simpler techniques for analyzing the effects of circumferential inlet distortion. Stenning concluded that it is impossible to achieve a complete success in calculating the distorted performance and distortion attenuation of an axial compressor even though the use of high speed computers has greatly improved the accuracy of performance analysis somewhat. However, as an attempt, it is necessary to study the distorted performance and distortion attenuation of an axial compressor even though a complete success may not be achievable. As a first approach, the methods that use the known undistorted performance characteristic as a starting point to predict the behavior of the compressor with distortion are developed with a much prospect for success. A number of investigators ([2], [3], [5], [6], [10]) have developed models for response to circumferential distortion which use the undistorted compressor performance or stage characteristics to predict the behavior with distortion. One novel real time correlation scheme for the detection of operations (both steady state and transient) near to the stability boundary of a compressor has recently been studied at the Georgia Institute of Technology [9]. The information provided by the correlation scheme is stochastic in nature to facilitate controller design [4]. The scheme has been linked to the engine fuel controller and was used detect and avoid impending stall by modulating the transient fuel schedule during operability tests [1].

Kim et al. [8] also successfully calculated the qualitative trend of distorted performance and distortion attenuation of an axial compressor by using a simple integral method. This simple integral method was applied to describe the qualitative trend of distortion propagation in axial compressors. Kim et al. concluded that the two key parameters to control the distortion propagation are the ratio of drag-to-lift coefficients of the blade and the angle of flow. The integral method provides the useful physical insight and qualitative information about the performance of the axial compressor with inlet distortion. It is meaningful to make use of and develop this method further.

In this chapter, the integral method is confirmed and further improved. An investigation is progressed to understand the downstream flow feature with the inlet distortion, including the velocity and mass flow rate. A critical distortion line is presented to express the effect of two essential inlet parameters on the propagation of distortion. The critical distortion line is simple, efficient and complete expression to analyze the propagation of inlet distortion in axial compressor. Finally, the compressor critical performance and critical characteristic are also discussed.

1.2 Theoretical Formulation

Consider a two-dimensional inviscid flow through an axial compressor as shown in [8]. The computational domain is shown in Fig. 1.1. The flow is described by equations of continuity and motion:

$$\frac{\partial u}{\partial x} + \frac{\partial v}{\partial y} = 0 \tag{1.1}$$

$$\frac{\partial}{\partial x}(u^2) + \frac{\partial}{\partial y}(uv) + \frac{\partial}{\partial x}(\frac{p}{\rho}) = F_x \tag{1.2}$$

$$\frac{\partial}{\partial x}(uv) + \frac{\partial}{\partial y}(v^2) + \frac{\partial}{\partial y}(\frac{p}{\rho}) = F_y \tag{1.3}$$

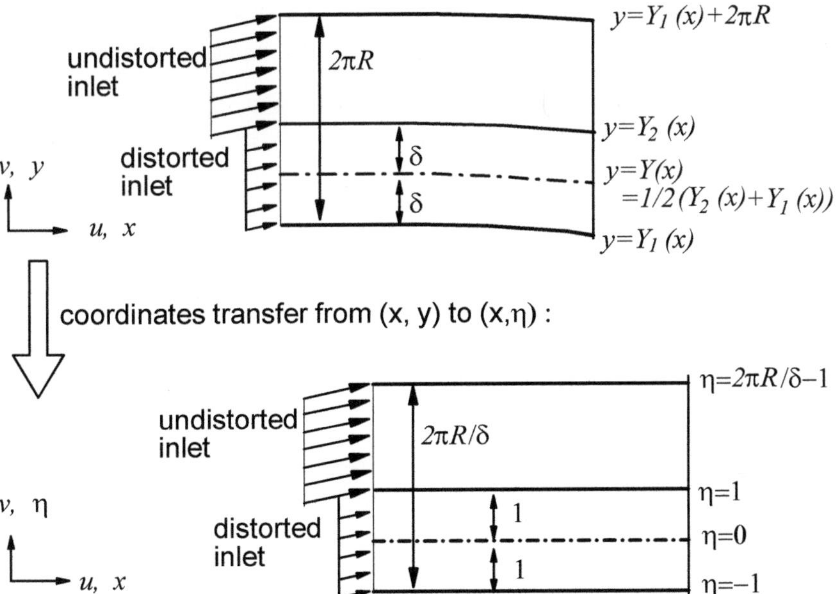

Fig. 1.1. A schematic of coordinates transfer of calculation domain

The coordinate system is transferred from (x, y) to (x, η) by:

$$\eta = \frac{y - Y(x)}{\delta} \tag{1.4}$$

here $y = Y(x)$ is a symmetric line of distorted region, and δ is a half height of distorted region.

If the static pressure takes circumferentially uniform, then:

$$\frac{p}{\rho} \equiv \frac{p}{\rho}(x) \tag{1.5}$$

Before integrating the x- and y- momentum equations, and mass conservation equation in distorted region, undistorted region and overall region, respectively, some assumptions and parameters definitions are given for velocity, pressure and forces.

1.2.1 Velocity and Pressure

Assume the inlet velocity has a windward angle of θ_0, then

$$\bar{\gamma} = tan\,\theta_0 = V_0/U_0 \tag{1.6}$$

where U_0 and V_0 are the x- and y- components of inlet velocity, respectively. α and β are the x- and y-velocity increments in the distorted inlet region, and α_0 and β_0 are the x- and y-velocity increments in the undistorted inlet region, respectively.

In distorted region:

$$u = \alpha U_0 \tag{1.7}$$

$$v = \beta V_0 \tag{1.8}$$

In undistorted region:

$$u = \alpha_0 U_0 \tag{1.9}$$

$$v = \beta_0 V_0 \tag{1.10}$$

The non-dimensionalized velocity parameter (rotor speed) is $\omega R = 2V_0$. The pressure is non-dimensionalized by $\frac{1}{2}\rho(2V_0)^2$:

$$P = \frac{p}{1/2\,\rho(2V_0)^2} \tag{1.11}$$

1.2.2 Forces

As shown in Fig. 1.2, in integral equations, F_x and F_y denote the forces in the distorted inlet region, and $F_{x,0}$ and $F_{y,0}$ denote the forces in the undistorted inlet region, respectively.

$$F_{\perp}^{S} = k_L \frac{1}{2}\left(u^2 + v^2\right)\left(tan\,\theta - tan\,\theta^*\right) \tag{1.12}$$

$$F_{II}^{S} = k_D \frac{1}{2}\left(u^2 + v^2\right)\left(tan\,\theta - tan\,\theta^*\right)^2 \tag{1.13}$$

$$F_{\perp}^{R} = k_L \frac{1}{2}\left[u^2 + \left(\omega R - v\right)^2\right]\left(tan\,\psi - tan\,\psi^*\right) \tag{1.14}$$

$$F_{II}^{R} = k_D \frac{1}{2}\left[u^2 + \left(\omega R - v\right)^2\right]\left(tan\,\psi - tan\,\psi^*\right)^2 \tag{1.15}$$

The force components for a unit circumferential distance of the entire blade row are:

$$F_x = \frac{\lambda_R}{\lambda}\left(F_{\perp}^{R}\,sin\,\psi - F_{II}^{R}\,cos\,\psi\right) + \frac{\lambda_S}{\lambda}\left(F_{\perp}^{S}\,sin\,\theta - F_{II}^{S}\,cos\,\theta\right) \tag{1.16}$$

$$F_y = \frac{\lambda_R}{\lambda}\left(F_{\perp}^{R}\,cos\,\psi + F_{II}^{R}\,sin\,\psi\right) + \frac{\lambda_S}{\lambda}\left(-F_{\perp}^{S}\,cos\,\theta - F_{II}^{S}\,sin\,\theta\right) \tag{1.17}$$

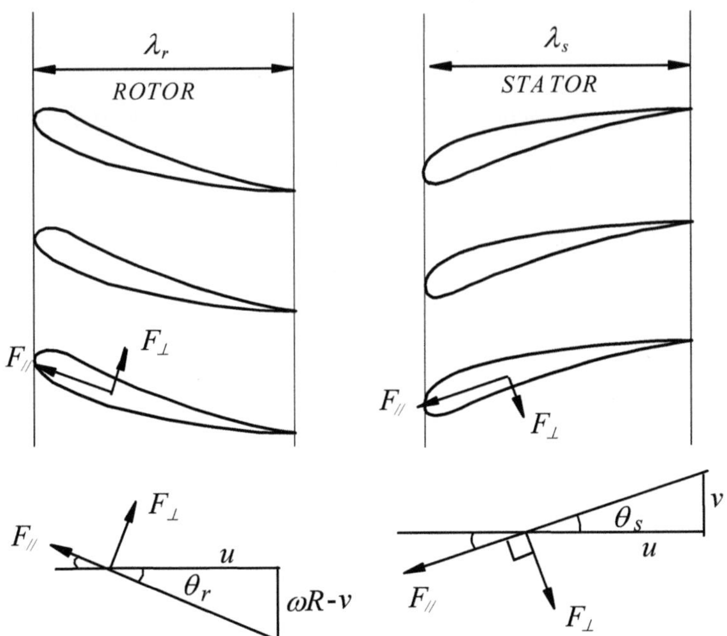

Fig. 1.2. The force diagram and velocities in the compressor stage

where the superscript S and R denote the stator and rotor, respectively; the λ_R/λ and λ_S/λ are generally functions of x ; k_L and k_D are lift and drag coefficients, respectively. The angles θ and ψ are the functions of the local velocity components and the parameter ω :

$$\tan\theta = v/u \tag{1.18}$$

$$\tan\psi = (\omega R - v)/u \tag{1.19}$$

Assuming $\omega R - V_0 = V_0$, in distorted region, by substituting (1.6), (1.7), (1.8), (1.18), (1.19) and (1.12), (1.13), (1.14), (1.15) into (1.16) and (1.17), we obtain:

$$
\begin{aligned}
F_x ={}& \frac{\lambda_R}{\lambda}\{\frac{k_L}{2}[u^2+(\omega R-v)^2](\tan\psi-\tan\psi^*)\sin\psi-\frac{k_D}{2}[u^2+(\omega R-v)^2](\tan\psi-\tan\psi^*)^2\cos\psi\}\\
&+\frac{\lambda_S}{\lambda}\{\frac{k_L}{2}(u^2+v^2)(\tan\theta-\tan\theta^*)\sin\theta-\frac{k_D}{2}(u^2+v^2)(\tan\theta-\tan\theta^*)^2\cos\theta\}\\
={}&\frac{\lambda_R}{\lambda}\frac{k_L}{2}\Big[\alpha^2U_0^2+(2-\beta)^2V_0^2\Big][(\frac{2-\beta}{\alpha}\bar\gamma-\tan\psi^*)\frac{2-\beta}{\alpha}\bar\gamma\\
&-\frac{k_D}{k_L}(\frac{2-\beta}{\alpha}\bar\gamma-\tan\psi^*)^2]\frac{\alpha}{\sqrt{\alpha^2+(2-\beta)^2\bar\gamma^2}}\\
&+\frac{\lambda_S}{\lambda}\frac{k_L}{2}(\alpha^2U_0^2+\beta^2V_0^2)[(\frac{\beta}{\alpha}\bar\gamma-\tan\theta^*)\frac{\beta}{\alpha}\bar\gamma-\frac{k_D}{k_L}(\frac{\beta}{\alpha}\bar\gamma-\tan\theta^*)^2]\frac{\alpha}{\sqrt{\alpha^2+\beta^2\bar\gamma^2}}
\end{aligned}
\tag{1.20}
$$

$$
\begin{aligned}
F_y ={}& \frac{\lambda_R}{\lambda}\{\frac{k_L}{2}[u^2+(\omega R-v)^2](\tan\psi-\tan\psi^*)\cos\psi+\frac{k_D}{2}[u^2+(\omega R-v)^2](\tan\psi-\tan\psi^*)^2\sin\psi\}\\
&+\frac{\lambda_S}{\lambda}\{-\frac{k_L}{2}(u^2+v^2)(\tan\theta-\tan\theta^*)\cos\theta-\frac{k_D}{2}(u^2+v^2)(\tan\theta-\tan\theta^*)^2\sin\theta\}\\
={}&\frac{\lambda_R}{\lambda}\frac{k_L}{2}\Big[\alpha^2U_0^2+(2-\beta)^2V_0^2\Big][(\frac{2-\beta}{\alpha}\bar\gamma-\tan\psi^*)\\
&+\frac{k_D}{k_L}(\frac{2-\beta}{\alpha}\bar\gamma-\tan\psi^*)^2\frac{2-\beta}{\alpha}\bar\gamma]\frac{\alpha}{\sqrt{\alpha^2+(2-\beta)^2\bar\gamma^2}}\\
&-\frac{\lambda_S}{\lambda}\frac{k_L}{2}(\alpha^2U_0^2+\beta^2V_0^2)[(\frac{\beta}{\alpha}\bar\gamma-\tan\theta^*)+\frac{k_D}{k_L}(\frac{\beta}{\alpha}\bar\gamma-\tan\theta^*)^2\frac{\beta}{\alpha}\bar\gamma]\frac{\alpha}{\sqrt{\alpha^2+\beta^2\bar\gamma^2}}
\end{aligned}
\tag{1.21}
$$

In undistorted region, by substituting (1.6), (1.9), (1.10), (1.20), (1.21) and (1.12), (1.13), (1.14), (1.15) into (1.16) and (1.17), using similar procedure as in the distorted region, we obtain:

$$F_{x,0} = \frac{\lambda_R}{\lambda}\frac{k_L}{2}\left[\alpha_0^2 U_0^2 + (2-\beta_0)^2 V_0^2\right]\left[(\frac{2-\beta_0}{\alpha_0}\bar{\gamma} - tan\psi^*)\frac{2-\beta_0}{\alpha_0}\bar{\gamma} - \right.$$
$$\left.\frac{k_D}{k_L}(\frac{2-\beta_0}{\alpha_0}\bar{\gamma} - tan\psi^*)^2\right]\frac{\alpha_0}{\sqrt{\alpha_0^2 + (2-\beta_0)^2\bar{\gamma}^2}} +$$
$$\frac{\lambda_S}{\lambda}\frac{k_L}{2}(\alpha_0^2 U_0^2 + \beta_0^2 V_0^2)\left[(\frac{\beta_0}{\alpha_0}\bar{\gamma} - tan\theta^*)\frac{\beta_0}{\alpha_0}\bar{\gamma} - \frac{k_D}{k_L}(\frac{\beta_0}{\alpha_0}\bar{\gamma} - tan\theta^*)^2\right]\frac{\alpha_0}{\sqrt{\alpha_0^2 + \beta_0^2\bar{\gamma}^2}}$$

$$(1.22)$$

$$F_{y,0} = \frac{\lambda_R}{\lambda}\frac{k_L}{2}\left[\alpha_0^2 U_0^2 + (2-\beta_0)^2 V_0^2\right]\left[(\frac{2-\beta_0}{\alpha_0}\bar{\gamma} - tan\psi^*) + \right.$$
$$\left.\frac{k_D}{k_L}(\frac{2-\beta_0}{\alpha_0}\bar{\gamma} - tan\psi^*)^2\frac{2-\beta_0}{\alpha_0}\bar{\gamma}\right]\frac{\alpha_0}{\sqrt{\alpha_0^2 + (2-\beta_0)^2\bar{\gamma}^2}} -$$
$$\frac{\lambda_S}{\lambda}\frac{k_L}{2}(\alpha_0^2 U_0^2 + \beta_0^2 V_0^2)\left[(\frac{\beta_0}{\alpha_0}\bar{\gamma} - tan\theta^*) + \frac{k_D}{k_L}(\frac{\beta_0}{\alpha_0}\bar{\gamma} - tan\theta^*)^2\frac{\beta_0}{\alpha_0}\bar{\gamma}\right]\frac{\alpha_0}{\sqrt{\alpha_0^2 + \beta_0^2\bar{\gamma}^2}}$$

$$(1.23)$$

Now, the (1.1), (1.2), (1.3) can be integrated in both distorted region and undistorted region.

1.2.3 Distorted Region

Accounting for the fact that the boundaries are streamlines, [8] developed the integral expressions as:

$$\delta \int_{-1}^{1} u d\eta = const. \tag{1.24}$$

$$\frac{d}{dx}\left[\delta \int_{-1}^{1} u^2 d\eta\right] + \int_{-1}^{1}\left[\delta\frac{d}{dx} - (Y' + \delta'\eta)\frac{\partial}{\partial\eta}\right]\frac{p}{\rho}d\eta = \delta \int_{-1}^{1} F_x d\eta \tag{1.25}$$

$$\frac{d}{dx}\left[\delta \int_{-1}^{1} uv d\eta\right] + \frac{p}{\rho}(x,1) - \frac{p}{\rho}(x,-1) = \delta \int_{-1}^{1} F_y d\eta \tag{1.26}$$

From (1.24) and (1.7), we can obtain a constant $\delta\alpha$. We introduce a constant K_1, and set:

$$\frac{\delta\alpha}{\pi R} \equiv K_1 \tag{1.27}$$

Then from (1.25):

$$\frac{d}{dx}\left[\delta \int_{-1}^{1} \alpha^2 U_0^2 d\eta\right] + \delta \int_{-1}^{1}\frac{d}{dx}(\frac{p}{\rho})d\eta\right] = \delta \int_{-1}^{1} F_x d\eta \tag{1.28}$$

Using (1.6), (1.11) and (1.27) and from (1.25) and (1.5), we can obtain:

$$\alpha \frac{d\alpha}{dx} + 2\overline{\gamma}^2 \frac{dP}{dx} = \frac{1}{U_0^2} F_x \tag{1.29}$$

Similarly, the y –momentum equation can be deducted from (1.26) plus the assumption of constant circumferential static pressure:

$$\frac{d}{dx}[\delta \int_{-1}^{1} \alpha \beta U_0 V_0 d\eta] = \delta \int_{-1}^{1} F_y d\eta \tag{1.30}$$

and then:

$$\alpha \frac{d\beta}{dx} = \frac{1}{\overline{\gamma}}(\frac{F_y}{U_0^2}) \tag{1.31}$$

1.2.4 Undistorted Region

Similarly, in undistorted region, integrating the (1.1), (1.2), (1.3) in $[1, \frac{2\pi R}{\delta}-1]$, we obtain:

$$\delta \int_{1}^{\frac{2\pi R}{\delta}-1} u d\eta = const. \tag{1.32}$$

$$\frac{d}{dx}[\delta \int_{1}^{\frac{2\pi R}{\delta}-1} u^2 d\eta]+ \int_{1}^{\frac{2\pi R}{\delta}-1}[\delta \frac{d}{dx} - (Y' + \delta'\eta)\frac{\partial}{\partial \eta}]\frac{p}{\rho} d\eta = \delta \int_{1}^{\frac{2\pi R}{\delta}-1} F_{x,0} d\eta \tag{1.33}$$

$$\frac{d}{dx}[\delta \int_{1}^{\frac{2\pi R}{\delta}-1} uv d\eta]+ \frac{p}{\rho}(x, \frac{2\pi R}{\delta}-1) - \frac{p}{\rho}(x, 1) = \delta \int_{1}^{\frac{2\pi R}{\delta}-1} F_{y,0} d\eta \tag{1.34}$$

From (1.33):

$$\frac{d}{dx}[\delta \int_{1}^{\frac{2\pi R}{\delta}-1} \alpha_0^2 U_0^2 d\eta]+ \delta \int_{1}^{\frac{2\pi R}{\delta}-1} \frac{d}{dx}(\frac{p}{\rho}) d\eta] = \delta \int_{1}^{\frac{2\pi R}{\delta}-1} F_{x,0} d\eta \tag{1.35}$$

or:

$$\frac{d}{dx}(\alpha_0^2) + 2\overline{\gamma}^2 \frac{dP}{dx} = \frac{1}{U_0^2} F_{x,0} \tag{1.36}$$

Similarly, for the y –momentum equation:

$$\frac{d}{dx}[\delta \int_{1}^{\frac{2\pi R}{\delta}-1} \alpha_0 \beta_0 U_0 V_0 d\eta] = \delta \int_{1}^{\frac{2\pi R}{\delta}-1} F_{y,0} d\eta \tag{1.37}$$

or:

$$\frac{d}{dx}(\alpha_0\beta_0) = \frac{1}{\overline{\gamma}}(\frac{F_{y,0}}{U_0^2}) \tag{1.38}$$

1.2.5 Entire Region

Integrating the overall conservation:

$$\delta\int_{-1}^{1}\alpha U_0\,d\eta + \delta\int_{1}^{\frac{2\pi R}{\delta}-1}\alpha_0 U_0\,d\eta \equiv const. \tag{1.39}$$

yield:

$$\left[2\delta\,\alpha U_0 + (2\pi R - 2\delta)(\alpha_0 U_0)\right]/(2\pi R U_0) \equiv K_0 \tag{1.40}$$

Using (1.27):

$$K_1 + (1 + K_1/\alpha)\alpha_0 = K_0 \tag{1.41}$$

Rearranged as:

$$\alpha_0 = \frac{\alpha(K_0 - K_1)}{\alpha - K_1} \tag{1.42}$$

The overall x- momentum equation is:

$$\frac{d}{dx}\delta\left[\int_{-1}^{1}\alpha^2 U_0^2\,d\eta + \int_{1}^{\frac{2\pi R}{\delta}-1}\alpha_0^2 U_0^2\,d\eta\right] + \delta\left[\int_{-1}^{1}\frac{2\pi R}{\delta}-1\frac{d}{dx}(\frac{p}{\rho})d\eta\right] = \delta\left[\int_{-1}^{1}F_x\,d\eta + \int_{1}^{\frac{2\pi R}{\delta}-1}F_{x,0}\,d\eta\right] \tag{1.43}$$

Integrated as:

$$\alpha\frac{d\alpha}{dx} - \frac{d\alpha_0^2}{dx} = \frac{1}{U_0^2}(F_x - F_{x,0}) \tag{1.44}$$

Differentiating (1.42) results in:

$$\frac{d\alpha_0}{dx} = K_2\frac{d\alpha}{dx} \tag{1.45}$$

Where:

$$K_2 = \frac{K_1(K_1 - K_0)}{(\alpha - K_1)^2} \tag{1.46}$$

Using (1.42) and (1.45):

$$\frac{d\alpha_0^2}{dx} = 2\alpha_0\frac{d\alpha_0}{dx} = -\frac{2K_1(K_1-K_0)^2}{(\alpha-K_1)^3}\alpha\frac{d\alpha}{dx} \tag{1.47}$$

The left-hand-side of (1.44) becomes:

$$\alpha\frac{d\alpha}{dx} - \frac{d\alpha_0}{dx}^2 = K_3\alpha\frac{d\alpha}{dx} \tag{1.48}$$

where:

$$K_3 = 1 + \frac{2K_1(K_1-K_0)^2}{(\alpha-K_1)^3} \tag{1.49}$$

Substituting above equation into (1.44), yield:

$$\alpha\frac{d\alpha}{dx} = \frac{1}{K_3 U_0^2}(F_x - F_{x,0}) \tag{1.50}$$

1.2.6 Integral Equations

From above integrated results, we can find out five ordinary differential equations which describe the progress of the distorted and undistorted regions, as well as the progress of pressure as flow moves downstream in the compressor.

The first one is (1.50), the second one is (1.31), and the third one is (1.45).

From the (1.45), we obtain:

$$\alpha_0\frac{d\beta_0}{dx} = \frac{1}{\gamma}(\frac{F_{y,0}}{U_0^2}) - \beta_0\frac{d\alpha_0}{dx} \tag{1.51}$$

Then, using (1.45), we obtain the fourth integral equation. From (1.29), the fifth integral equation can be derived.

Putting all of the five integral equations together, there are:

$$\alpha\frac{d\alpha}{dx} = \frac{1}{K_3}(\frac{F_x - F_{x,0}}{U_0^2}) \tag{1.52}$$

$$\alpha\frac{d\beta}{dx} = \frac{1}{\gamma}(\frac{F_y}{U_0^2}) \tag{1.53}$$

$$\frac{d\alpha_0}{dx} = K_2\frac{d\alpha}{dx} \tag{1.54}$$

$$\alpha_0 \frac{d\beta_0}{dx} = \frac{1}{\bar{\gamma}}\left(\frac{F_{y,0}}{U_0^2}\right) - \beta_0 K_2 \frac{d\alpha}{dx} \qquad (1.55)$$

$$2\bar{\gamma}^2 \frac{dP}{dx} = \frac{F_x}{U_0^2} - \alpha \frac{d\alpha}{dx} \qquad (1.56)$$

These five differential equations contain terms obtained by the integration of differential expressions of the distortion parameters, following up with their previous name [8], we term them as integral equations here.

1.3 Numerical Method

1.3.1 Equations

The integral (1.52, 1.53, 1.54, 1.55, 1.56) are solved numerical using the 4th-order Runge-Kutta method. The integral equations are rewritten as:

$$\frac{d\alpha}{dx} = \frac{1}{\alpha K_3}\left(\frac{F_x - F_{x,0}}{U_0^2}\right) = f_1(x,\alpha,\beta,\alpha_0,\beta_0,P) \qquad (1.57)$$

$$\frac{d\beta}{dx} = \frac{1}{\alpha\bar{\gamma}}\left(\frac{F_y}{U_0^2}\right) = f_2(x,\alpha,\beta,\alpha_0,\beta_0,P) \qquad 1.58)$$

$$\frac{d\alpha_0}{dx} = \frac{K_2}{\alpha K_3}\left(\frac{F_x - F_{x,0}}{U_0^2}\right) = f_3(x,\alpha,\beta,\alpha_0,\beta_0,P) \qquad (1.59)$$

$$\frac{d\beta_0}{dx} = \frac{1}{\alpha_0\bar{\gamma}}\left(\frac{F_{y,0}}{U_0^2}\right) - \frac{\beta_0 K_2}{\alpha_0 \alpha K_3}\frac{F_x - F_{x,0}}{U_0^2} = f_4(x,\alpha,\beta,\alpha_0,\beta_0,P) \qquad (1.60)$$

$$\frac{dP}{dx} = \left(\frac{F_x}{U_0^2} - \alpha\frac{d\alpha}{dx}\right)\frac{1}{2\bar{\gamma}^2} = f_5(x,\alpha,\beta,\alpha_0,\beta_0,P) \qquad (1.61)$$

1.3.2 4th-order Runge-Kutta Equations

$$\alpha_{i+1} = \alpha_i + \frac{1}{6}\left(K_{11} + 2K_{21} + 2K_{31} + K_{41}\right) \qquad (1.62)$$

$$\beta_{i+1} = \beta_i + \frac{1}{6}\left(K_{12} + 2K_{22} + 2K_{32} + K_{42}\right) \qquad (1.63)$$

$$\alpha_{0\,i+1} = \alpha_{0\,i} + \frac{1}{6}\left(K_{13} + 2K_{23} + 2K_{33} + K_{43}\right) \qquad (1.64)$$

$$\beta_{0\,i+1} = \beta_{0\,i} + \frac{1}{6}\left(K_{14} + 2K_{24} + 2K_{34} + K_{44}\right) \qquad (1.65)$$

$$P_{i+1} = P_i + \frac{1}{6}\left(K_{15} + 2K_{25} + 2K_{35} + K_{45}\right) \qquad (1.66)$$

where, the iterative variables can be calculated by:

$$K_{11} = \Delta x \, f_1(x_i, \alpha_i, \beta_i, \alpha_{0_i}, \beta_{0_i}, P_i) \tag{1.67}$$

$$K_{12} = \Delta x \, f_2(x_i, \alpha_i, \beta_i, \alpha_{0_i}, \beta_{0_i}, P_i) \tag{1.68}$$

$$K_{13} = \Delta x \, f_3(x_i, \alpha_i, \beta_i, \alpha_{0_i}, \beta_{0_i}, P_i) \tag{1.69}$$

$$K_{14} = \Delta x \, f_4(x_i, \alpha_i, \beta_i, \alpha_{0_i}, \beta_{0_i}, P_i) \tag{1.70}$$

$$K_{15} = \Delta x \, f_5(x_i, \alpha_i, \beta_i, \alpha_{0_i}, \beta_{0_i}, P_i) \tag{1.71}$$

$$K_{21} = \Delta x \, f_1(x_i + \frac{\Delta x}{2}, \alpha_i + \frac{K_{11}}{2}, \beta_i + \frac{K_{12}}{2}, \alpha_{0_i} + \frac{K_{13}}{2}, \beta_{0_i} + \frac{K_{14}}{2}, P_i + \frac{K_{15}}{2}) \tag{1.72}$$

$$K_{22} = \Delta x \, f_2(x_i + \frac{\Delta x}{2}, \alpha_i + \frac{K_{11}}{2}, \beta_i + \frac{K_{12}}{2}, \alpha_{0_i} + \frac{K_{13}}{2}, \beta_{0_i} + \frac{K_{14}}{2}, P_i + \frac{K_{15}}{2}) \tag{1.73}$$

$$K_{23} = \Delta x \, f_3(x_i + \frac{\Delta x}{2}, \alpha_i + \frac{K_{11}}{2}, \beta_i + \frac{K_{12}}{2}, \alpha_{0_i} + \frac{K_{13}}{2}, \beta_{0_i} + \frac{K_{14}}{2}, P_i + \frac{K_{15}}{2}) \tag{1.74}$$

$$K_{24} = \Delta x \, f_4(x_i + \frac{\Delta x}{2}, \alpha_i + \frac{K_{11}}{2}, \beta_i + \frac{K_{12}}{2}, \alpha_{0_i} + \frac{K_{13}}{2}, \beta_{0_i} + \frac{K_{14}}{2}, P_i + \frac{K_{15}}{2}) \tag{1.75}$$

$$K_{25} = \Delta x \, f_5(x_i + \frac{\Delta x}{2}, \alpha_i + \frac{K_{11}}{2}, \beta_i + \frac{K_{12}}{2}, \alpha_{0_i} + \frac{K_{13}}{2}, \beta_{0_i} + \frac{K_{14}}{2}, P_i + \frac{K_{15}}{2}) \tag{1.76}$$

$$K_{31} = \Delta x \, f_1(x_i + \frac{\Delta x}{2}, \alpha_i + \frac{K_{21}}{2}, \beta_i + \frac{K_{22}}{2}, \alpha_{0_i} + \frac{K_{23}}{2}, \beta_{0_i} + \frac{K_{24}}{2}, P_i + \frac{K_{25}}{2}) \tag{1.77}$$

$$K_{32} = \Delta x \, f_2(x_i + \frac{\Delta x}{2}, \alpha_i + \frac{K_{21}}{2}, \beta_i + \frac{K_{22}}{2}, \alpha_{0_i} + \frac{K_{23}}{2}, \beta_{0_i} + \frac{K_{24}}{2}, P_i + \frac{K_{25}}{2}) \tag{1.78}$$

$$K_{33} = \Delta x \, f_3(x_i + \frac{\Delta x}{2}, \alpha_i + \frac{K_{21}}{2}, \beta_i + \frac{K_{22}}{2}, \alpha_{0_i} + \frac{K_{23}}{2}, \beta_{0_i} + \frac{K_{24}}{2}, P_i + \frac{K_{25}}{2}) \tag{1.79}$$

$$K_{34} = \Delta x \, f_4(x_i + \frac{\Delta x}{2}, \alpha_i + \frac{K_{21}}{2}, \beta_i + \frac{K_{22}}{2}, \alpha_{0_i} + \frac{K_{23}}{2}, \beta_{0_i} + \frac{K_{24}}{2}, P_i + \frac{K_{25}}{2}) \tag{1.80}$$

$$K_{35} = \Delta x \, f_5(x_i + \frac{\Delta x}{2}, \alpha_i + \frac{K_{21}}{2}, \beta_i + \frac{K_{22}}{2}, \alpha_{0_i} + \frac{K_{23}}{2}, \beta_{0_i} + \frac{K_{24}}{2}, P_i + \frac{K_{25}}{2}) \tag{1.81}$$

$$K_{41} = \Delta x \, f_1(x_i + \Delta x, \alpha_i + K_{31}, \beta_i + K_{32}, \alpha_{0_i} + K_{33}, \beta_{0_i} + K_{34}, P_i + K_{35}) \tag{1.82}$$

$$K_{42} = \Delta x \, f_2(x_i + \Delta x, \alpha_i + K_{31}, \beta_i + K_{32}, \alpha_{0_i} + K_{33}, \beta_{0_i} + K_{34}, P_i + K_{35}) \tag{1.83}$$

$$K_{43} = \Delta x \, f_3(x_i + \Delta x, \alpha_i + K_{31}, \beta_i + K_{32}, \alpha_{0_i} + K_{33}, \beta_{0_i} + K_{34}, P_i + K_{35}) \tag{1.84}$$

$$K_{44} = \Delta x \, f_4(x_i + \Delta x, \alpha_i + K_{31}, \beta_i + K_{32}, \alpha_{0_i} + K_{33}, \beta_{0_i} + K_{34}, P_i + K_{35}) \tag{1.85}$$

$$K_{45} = \Delta x \, f_5(x_i + \Delta x, \alpha_i + K_{31}, \beta_i + K_{32}, \alpha_{0_i} + K_{33}, \beta_{0_i} + K_{34}, P_i + K_{35}) \tag{1.86}$$

1.4 Results and Discussion

In [8], for simplify the numerical calculation, the following assumptions was made:

(1) $\lambda_R = \lambda_S$

(2) $\theta^* = \psi^*$

(3) $\omega R - V_0 = V_0$

(4) $\alpha_{0\ x=0} = 1$

Therefore, the expressions of the forces, (1.20, 1.21, 1.22, 1.23), can be simplified as:

$$
F_x = \frac{k_L}{4}U_0^{\,2}\left[\alpha^2 + (2-\beta)^2\,\bar\gamma^2\right]\left(\frac{2-\beta}{\alpha}\bar\gamma - \zeta\right)\left[\frac{2-\beta}{\alpha}\bar\gamma\left(1-\frac{k_D}{k_L}\right)+\frac{k_D}{k_L}\zeta\right]\frac{\alpha}{\sqrt{\alpha^2+(2-\beta)^2\bar\gamma^2}}
$$
$$
+\frac{k_L}{4}U_0^{\,2}(\alpha^2+\beta^2\bar\gamma^2)\left(\frac{\beta}{\alpha}\bar\gamma - \zeta\right)\left[\frac{\beta}{\alpha}\bar\gamma\left(1-\frac{k_D}{k_L}\right)+\frac{k_D}{k_L}\zeta\right]\frac{\alpha}{\sqrt{\alpha^2+\beta^2\bar\gamma^2}}
$$

$$\tag{1.87}$$

$$
F_y = \frac{k_L}{4}U_0^{\,2}\left[\alpha^2 + (2-\beta)^2\,\bar\gamma^2\right]\left(\frac{2-\beta}{\alpha}\bar\gamma - \zeta\right)\left[1+\frac{k_D}{k_L}\left(\frac{2-\beta}{\alpha}\bar\gamma - \zeta\right)\frac{2-\beta}{\alpha}\bar\gamma\right]\frac{\alpha}{\sqrt{\alpha^2+(2-\beta)^2\bar\gamma^2}}
$$
$$
-\frac{k_L}{4}U_0^{\,2}(\alpha^2+\beta^2\bar\gamma^2)\left(\frac{\beta}{\alpha}\bar\gamma - \zeta\right)\left[1+\frac{k_D}{k_L}\left(\frac{\beta}{\alpha}\bar\gamma - \zeta\right)\frac{\beta}{\alpha}\bar\gamma\right]\frac{\alpha}{\sqrt{\alpha^2+\beta^2\bar\gamma^2}}
$$

$$\tag{1.88}$$

$$
F_{x,0} = \frac{k_L}{4}U_0^2\left[\alpha_0^{\,2} + (2-\beta_0)^2\,\bar\gamma^2\right]\left(\frac{2-\beta_0}{\alpha_0}\bar\gamma - \zeta\right)\left[\frac{2-\beta_0}{\alpha_0}\bar\gamma\left(1-\frac{k_D}{k_L}\right)+\frac{k_D}{k_L}\zeta\right]\frac{\alpha_0}{\sqrt{\alpha_0^{\,2}+(2-\beta_0)^2\bar\gamma^2}}
$$
$$
+\frac{k_L}{4}U_0^{\,2}(\alpha_0^{\,2}+\beta_0^2\bar\gamma^2)\left(\frac{\beta_0}{\alpha_0}\bar\gamma - \zeta\right)\left[\frac{\beta_0}{\alpha_0}\bar\gamma\left(1-\frac{k_D}{k_L}\right)+\frac{k_D}{k_L}\zeta\right]\frac{\alpha_0}{\sqrt{\alpha_0^{\,2}+\beta_0^2\bar\gamma^2}}
$$

$$\tag{1.89}$$

$$
F_{y,0} = \frac{k_L}{4}U_0^{\,2}\left[\alpha_0^{\,2} + (2-\beta_0)^2\,\bar\gamma^2\right]\left(\frac{2-\beta_0}{\alpha_0}\bar\gamma - \zeta\right)\left[1+\frac{k_D}{k_L}\left(\frac{2-\beta_0}{\alpha_0}\bar\gamma - \zeta\right)\frac{2-\beta_0}{\alpha_0}\bar\gamma\right]\frac{\alpha_0}{\sqrt{\alpha_0^{\,2}+(2-\alpha_0)^2\bar\gamma^2}}
$$
$$
-\frac{k_L}{4}U_0^{\,2}(\alpha_0^{\,2}+\beta_0^2\bar\gamma^2)\left(\frac{\beta_0}{\alpha_0}\bar\gamma - \zeta\right)\left[1+\frac{k_D}{k_L}\left(\frac{\beta_0}{\alpha_0}\bar\gamma - \zeta\right)\frac{\beta_0}{\alpha_0}\bar\gamma\right]\frac{\alpha_0}{\sqrt{\alpha_0^{\,2}+\beta_0^2\bar\gamma^2}}
$$

$$\tag{1.90}$$

where, $\zeta = \tan\theta^* = \tan\psi^*$.

1.4.1 Previous Results

The distortion propagation is denoted by vertical extension of distorted flow, ξ :

$$
\xi(x) = \frac{\delta}{\pi R} = K_1/\alpha
$$

$$\tag{1.91}$$

The lift coefficient is taken as $k_L = 3.0$ here.

Kim et al. [8] presented five cases in their paper. These cases were produced by solving the integral equations, (1.52), (1.53) and (1.55). The conditions and result in these cases can be summarized as in Table 1.1. Here, k_D/k_L is the drag-to-lift ratio; $\alpha_{x=0} = \beta_{x=0}$ is the initial distortion, and, $\alpha_{x=0} = \alpha(0)$; $\beta_{x=0} = \beta(0)$.

There are two main questions about the calculations in [8]. Firstly, in their calculations, both (1.54) and (1.56) were not included and the results of variable ξ were presented. In fact, the calculation of variable ξ is followed by the calculation of the variables α, β, α_0 and β_0. However, Kim et al derived only three integral equations (1.52), (1.53) and (1.55). They must put one more condition such as $\alpha_0 = \beta_0$ in their calculation so that to work out the four parameters, although this was not mentioned in their paper. It is why there are differences between Kim's and present results, as shown in Fig. 1.3. By using complete integral equations and less assumption, present results are certainly more reasonable. Their assumption of $\alpha_{0\,x=0} = \beta_{0\,x=0}$ is not sufficient for numerical calculation. The lift coefficient is another question. The results strongly depended on the lift coefficient, k_L. On the contrary, in the normal range of the drag-to-lift ratio coefficient, the results appear less variation. Reference [8] gave only the drag-to-lift ratio coefficient, k_D/k_L. By means of numerical experiment, we found that their results might be obtained

Table 1.1 The calculating cases in previous paper [8]

case	conditions	results
1	Fix: $\theta_0 = 15^0$, $\theta^* = \psi^* = 10^0$, $k_D/k_L = 1$ Change: $\alpha_{x=0} = \beta_{x=0} = 0.3$, 0.5, 0.7	The effect of initial inlet distortion $\alpha_{x=01}$, $\beta_{x=0}$. (Fig. 1.3)
2	Fix: $\alpha_{x=0} = \beta_{x=0} = 0.7$, $\theta^* = \psi^* = 10^0$, $k_D/k_L = 1$ Change: $\theta_0 = 15^0$, $\theta_0 = 25^0$	The effect of inlet flow angle, θ_0. (Fig. 1.4)
3	Fix: $\theta_0 = 25^0$, $\alpha_{x=0} = \beta_{x=0} = 0.7$, $\theta^* = \psi^* = 10^0$ Change: $k_D/k_L = 0.80$, 0.85, 0.90, 0.95	The effect of drag-to-lift ratio, k_D/k_L. (Fig. 1.5)
4	Fix: $\theta_0 = 25^0$, $\theta^* = \psi^* = 10^0$, $k_D/k_L = 1$ Change: $\alpha_{x=0} = \beta_{x=0} = 0.7$, 0.95.	The effect of high inlet flow angle θ_0. (Fig. 1.6)
5	Fix: $\theta_0 = 25^0$, $\theta^* = \psi^* = 10^0$, $k_D/k_L = 1$ Change: $\alpha_{x=0} = \beta_{x=0} = 0.7$, 0.95. $x \to \infty$	The propagation of inlet distortion toward far downstream. (Fig. 1.6)

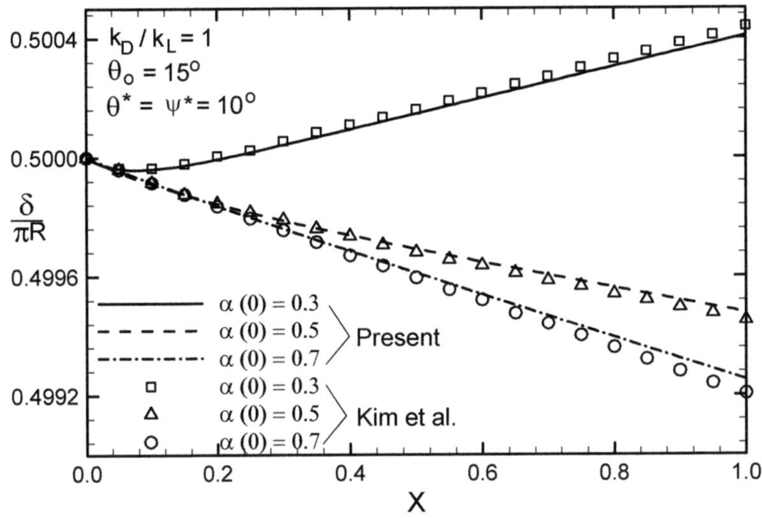

Fig. 1.3. The effect of the degree of the initial distortion

by assuming $k_D/k_L = 3.0$. For ease of comparison, $k_D/k_L = 3.0$ was used here too. Finally, the initial distortion, $\xi(0) = \dfrac{\delta}{\pi R}\Big|_{x=0}$, should be specified at a value of 0.5 according to their figures presented.

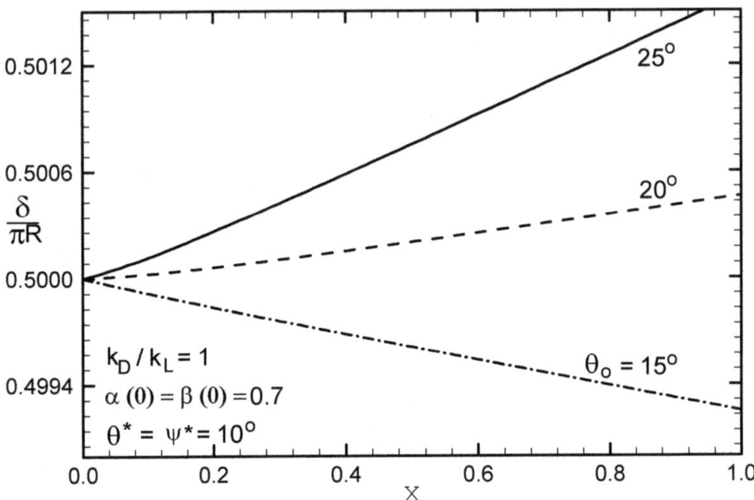

Fig. 1.4. The effect of the angle of flow

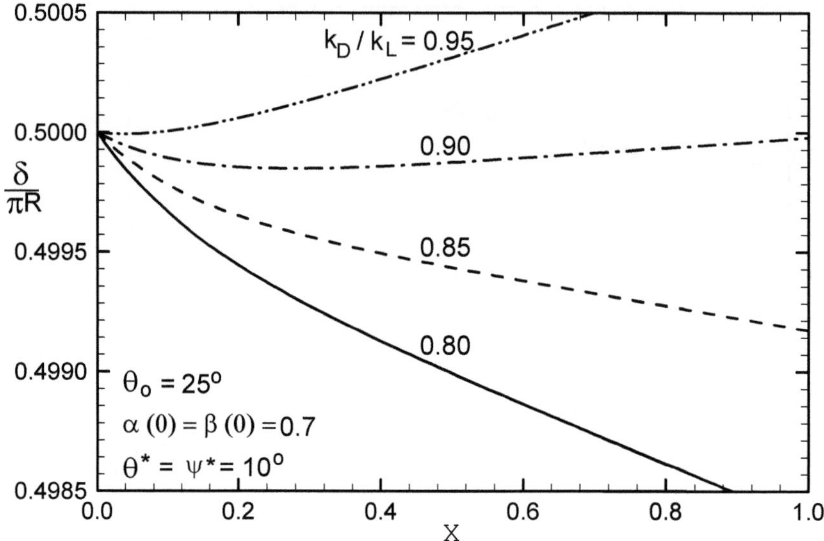

Fig. 1.5. The effect of the drag-to-lift ratio

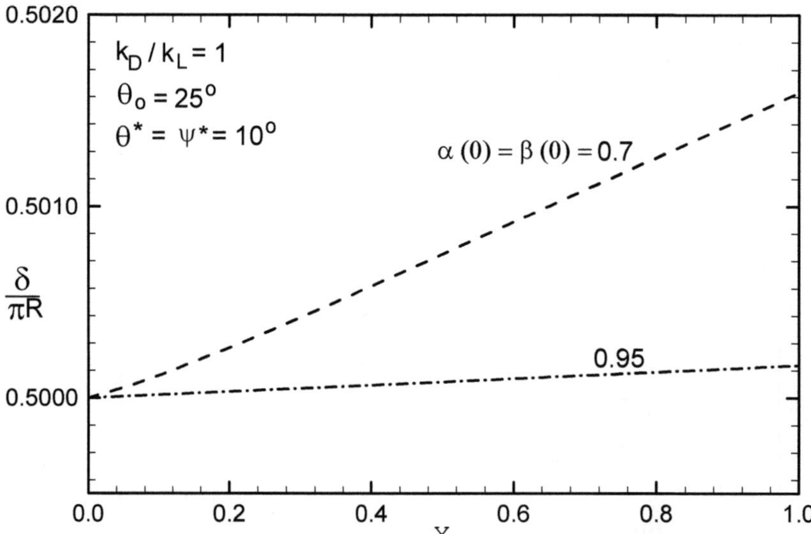

Fig. 1.6. The trend of distortion propagation at a high flow angle

Case 1, case 2 and case 4 can be combined to a single case of "critical distortion line" that to be discussed in the next section. Case 3 is not necessary because the interest here is distorted inlet propagation. The affecting parameters should involve only that of the inlet conditions. The compressor should be considered as a given geometric domain. As to the asymptotic behavior situation, case 5, there is much further work to do, which will be mentioned next also.

1.4.2 Further Asymptotic Behavior Results

The four variables in integral (1.52) to (1.55) have different asymptotic behaviors. Their asymptotic behaviors are calculated with the same conditions as in case 5. Fig. 1.7 indicates that the distortion grows in downstream direction. The axial component of velocity arrives its asymptote gradually: $x > 300$ for $\alpha_{x=0} = \beta_{x=0} = 0.7$, and even slower for larger inlet distortion. The asymptotic trend of axial velocity in the distorted region is contrary to that in undistorted regions. The value of α tends to a lower asymptote but α_0 tends to a higher one as shown in Fig. 1.8 and Fig. 1.9 respectively. These results present the status of the distorted region downstream in the case of the growing distortion. At the far downstream, the distorted region increases in its size, and becomes more explicit because the axial velocity decreases in distorted region, and increases in undistorted region.

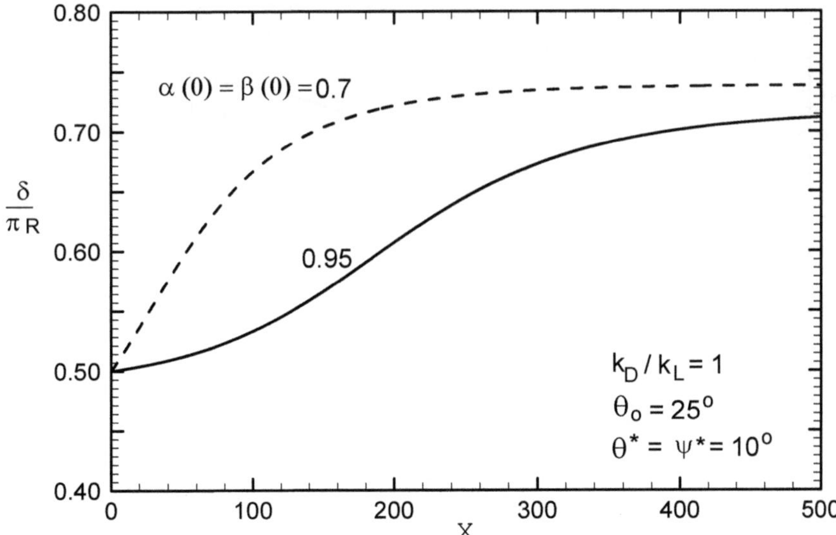

Fig. 1.7. The asymptotic behavior of the inlet distortion propagation

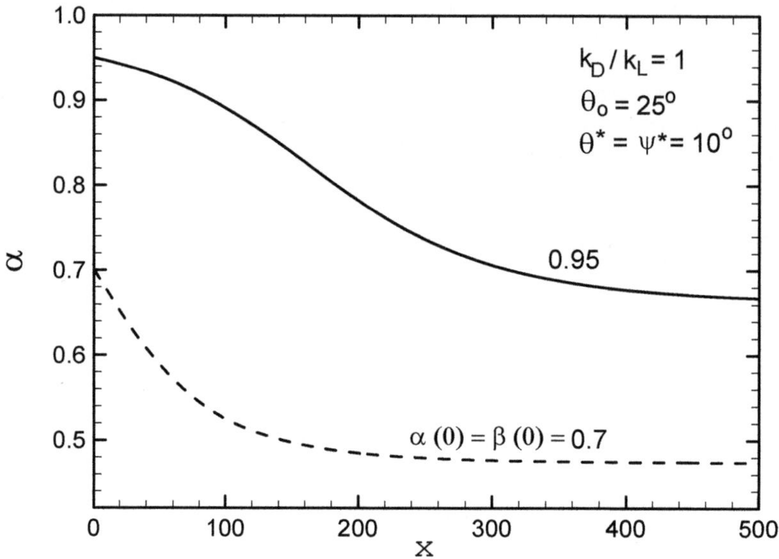

Fig. 1.8. The asymptotic behavior of α

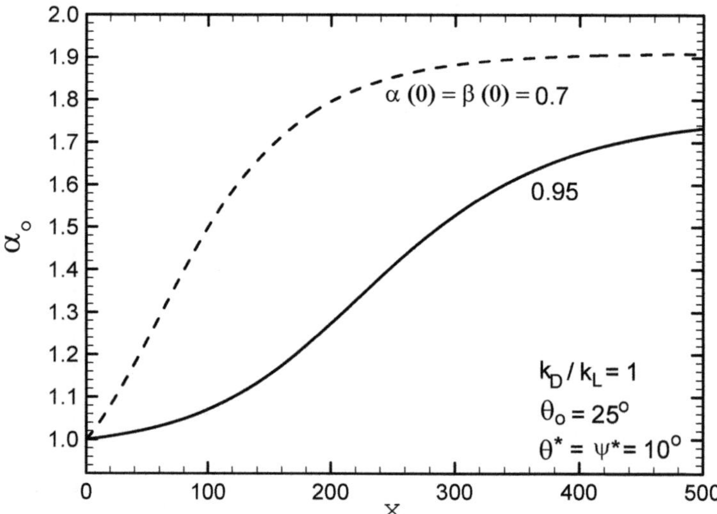

Fig. 1.9. The asymptotic behavior of α_0

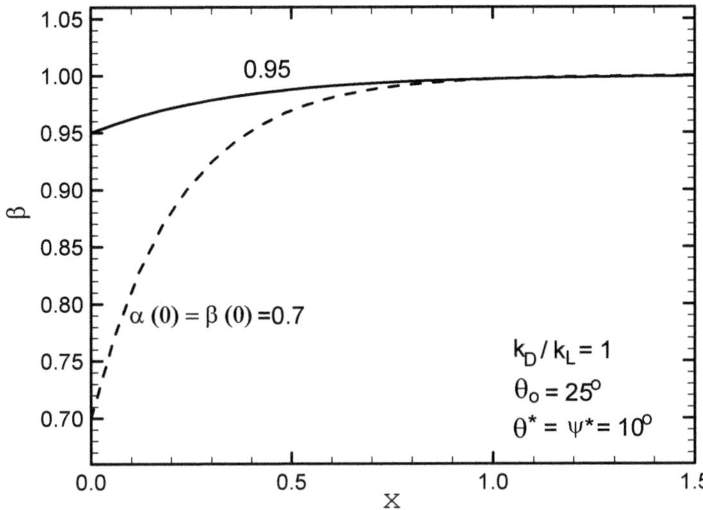

Fig. 1.10. The asymptotic behavior of β near the inlet

From (1.91), the size of distorted region $\xi(x)$ is inverse proportional to the velocity in distorted region. This relation explains the reason that axial velocity in distorted region has a contrary trend with the size of distorted region, $\xi(x)$, and the distortion region may propagate downstream with the growing of $\xi(x)$ when $d\alpha/dx < 0$.

The variation of normal velocity is relatively much smaller as shown in Fig. 1.10 and Fig. 1.11 respectively. The vertical component of velocity in distorted region grows up to its asymptote rapidly: $x < 1.0$, and the asymptote is equal to the vertical component of the inlet velocity in undistorted region. In other words, the value of β arrives its asymptote β_0 in a shorter distance of $\Delta x < 1.0$. Consequently, the velocities in the two regions have different directions downstream; even they have a same inlet flow angle, θ_0. The flow angle of velocity tends to increase in distorted region, and decrease in undistorted region in this inlet distortion growing case.

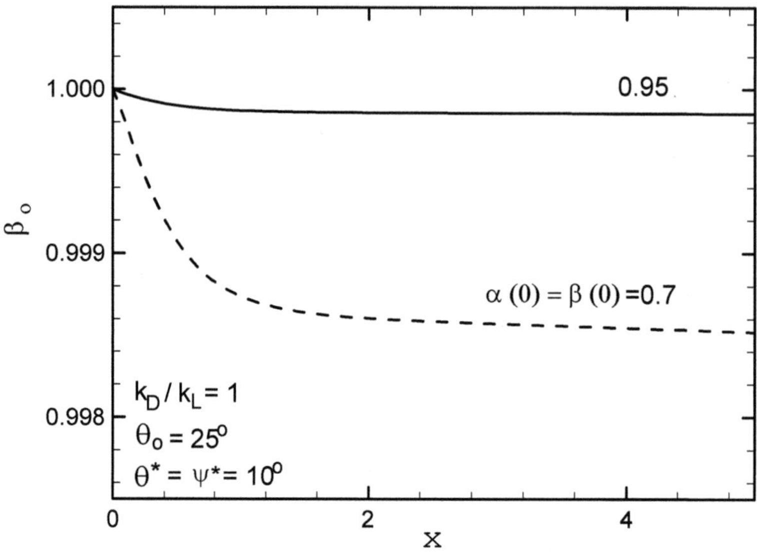

Fig. 1.11. The asymptotic behavior of β_0 near the inlet

1.4.3 Mass Flow Rate

From the overall mass conservation equation, the mass flow rate is:

$$\dot{m} = \rho\delta \int_{-1}^{1} \alpha U_0 d\eta + \rho\delta \int_{1}^{2\pi R/\delta - 1} \alpha U_0 d\eta \tag{1.92}$$

Using (1.91), we obtain:

$$\dot{m} = 2\rho U_0 \pi R \left[(\alpha - \alpha_0)\xi + 1 \right] \tag{1.93}$$

Non-dimensionalize the (1.93) using $2\pi R\rho(2V_0)$, we obtain:

$$\dot{M} = \left[(\alpha - \alpha_0)\xi + 1 \right] / (2\bar{\gamma}) \tag{1.94}$$

Fig. 1.12 shows the asymptotic trend of mass flow rate. In this case, all calculation conditions are the same as in case 5, which is an unstable case because its distortion grows downstream as shown in Fig. 1.12. In this unstable case, the mass flow rate decreases sharply. A smaller inlet distorted velocity induces a larger reduction in mass

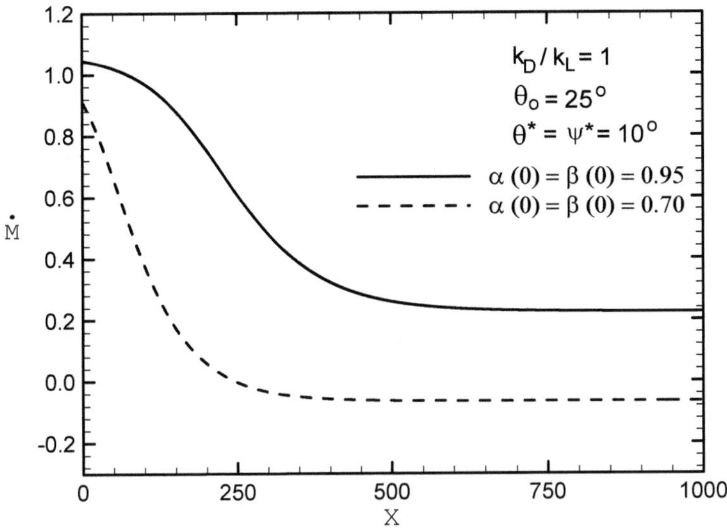

Fig. 1.12. The asymptotic behavior of mass flow rate

flow rate. Besides, a higher mass flow rate will arrive at a larger asymptote. This proportional relation between mass flow rate and inlet distorted velocity can also be seen from Fig. 1.13. In conclusion, a smaller inlet distorted velocity and larger flow angle may result in a low overall mass flow rate, even a back flow (Fig. 1.13).

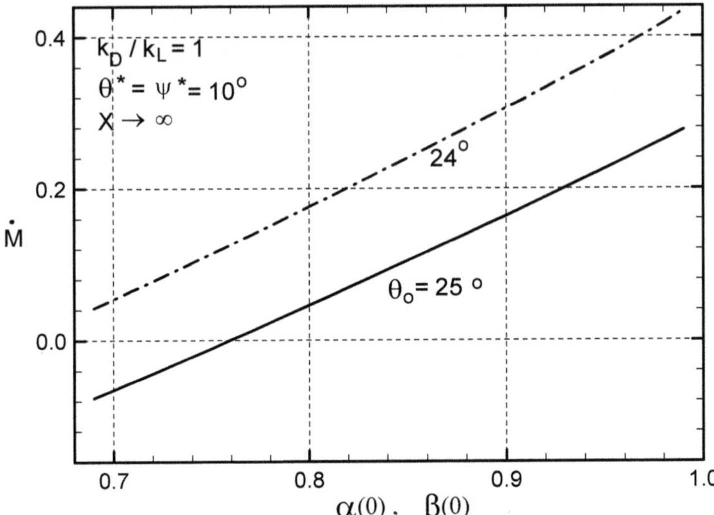

Fig. 1.13. The effects of degree of the initial distortion and the inlet flow angles on the mass flow rate

1.4.4 Critical Distortion Line

The results in previous paper [8], case 1, case 2 and case 4 expressed the distortion propagation with respect to different inlet flow angle and different inlet distorted velocity. These results are isolated and incomplete. A more suitable analysis way is combining the two parameters, inlet flow angle and inlet distorted velocity, into a single figure to form a curve, as shown in Fig. 1.14. We term the curve with $\Delta\xi = \xi_{x=L} - \xi_{x=0} = 0$ as "critical distortion line". Above the critical distortion line, $\Delta\xi = \xi_{x=L} - \xi_{x=0} > 0$, the distorted region will grow downstream and become unstable. On the contrary, below the critical distortion line, $\Delta\xi = \xi_{x=L} - \xi_{x=0} < 0$, the distorted region will reduce downstream and become stable. This critical distortion line is a more complete expression for inlet distortion propagation. The six numbering circles in Fig. 1.14 correspond to all results in case 1, case 2 and case 4 of Table 1.1, and their results are listed in Table 1.2.

It is very worthy to note that with different size of distortion region, the compressor performs with only one critical distortion line. In other words, the size of distortion region has no influence on the critical distortion line. Therefore, the proposed critical distortion line includes the principal effects at inlet.

Table 1.2 The corresponding relation between the numbering circles in Fig. 1.14 and cases calculated in previous paper [8]

circle no.	conditions	distortion status
1	case 1 $\alpha_{x=0} = \beta_{x=0} = 0.3$	growing.
2	case 1 $\alpha_{x=0} = \beta_{x=0} = 0.5$	lessening.
3	case 1 $\alpha_{x=0} = \beta_{x=0} = 0.7$	lessening.
3	case 2 $\theta_0 = 15^0$	lessening.
4	case 2 $\theta_0 = 20^0$	growing.
5	case 2 $\theta_0 = 25^0$	growing.
5	case 4 $\alpha_{x=0} = \beta_{x=0} = 0.7$	growing.
6	case 4 $\alpha_{x=0} = \beta_{x=0} = 0.95$	growing.

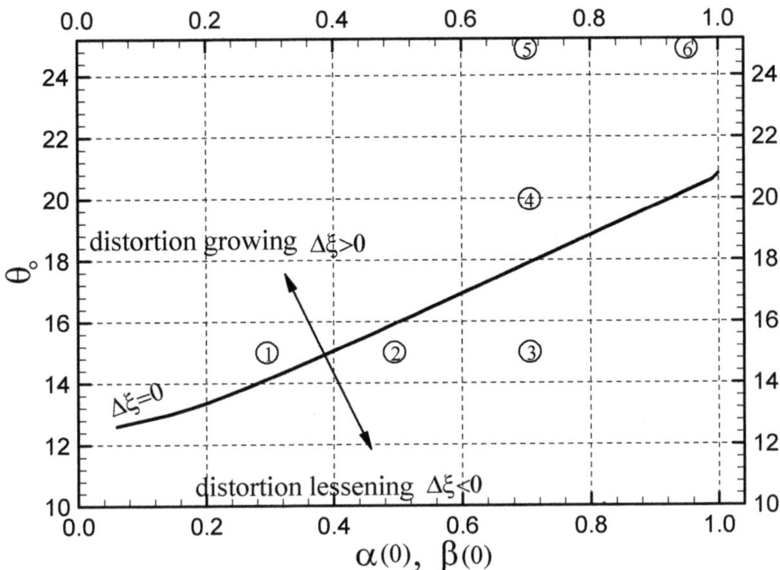

Fig. 1.14. Critical distortion line

From Fig. 1.14, the critical distortion line is nearly a straight line:

$$\theta_0 = A_1\alpha(0) + A_2 \tag{1.95}$$

The two constants can be found out from Fig. 1.14: $A_1 = 9.43$ and $A_2 = 11.266$.

In establishing the correlation between stall inception and inlet distortion, the critical distortion line is described mathematically using a distortion propagation factor according to (1.95):

$$\varphi = \frac{\theta_0}{A_1\alpha + A_2} \tag{1.96}$$

The region with $\varphi > 1$ denotes where the propagation will grow and the region with $\varphi < 1$ denotes where the propagation will decay.

1.4.5 Compressor Performance and Characteristic

A pressure profile can be obtained by solving (1.56) together with (1.52), (1.53), (1.54), (1.55). The inlet and outlet total pressure are written as:

$$\frac{P_{0\,out}}{P_{0\,in}} = \frac{P_2 + v_2\,/(2V_0\,)^2}{P_1 + v_1\,/(2V_0\,)^2} \qquad (1.97)$$

Here, the inlet air is at the ambient atmospheric condition. P_1 and P_2 are inlet and outlet non-dimensional pressure, v_1 and v_2 are inlet and outlet velocity without distortion, respectively.

The compressor performance corresponding to the critical distortion line as shown in Fig. 1.15 is termed as a compressor critical performance. By changing the size of inlet distortion region $\xi(0)$ from *0.5* to *0.0*, the different results of the compressor critical performance can be produced as shown in Fig. 1.16. The effect of the size of inlet distortion region on the compressor critical performance agrees with intuitive anticipation that larger size of inlet distortion region induces a decrease of total mass flux.

The non-dimensional velocity and pressure rise in Fig. 1.17 are defined as:

$$\phi = \frac{u}{2V_0} \qquad (1.98)$$

$$\psi_p = P_2 - P_1 \qquad (1.99)$$

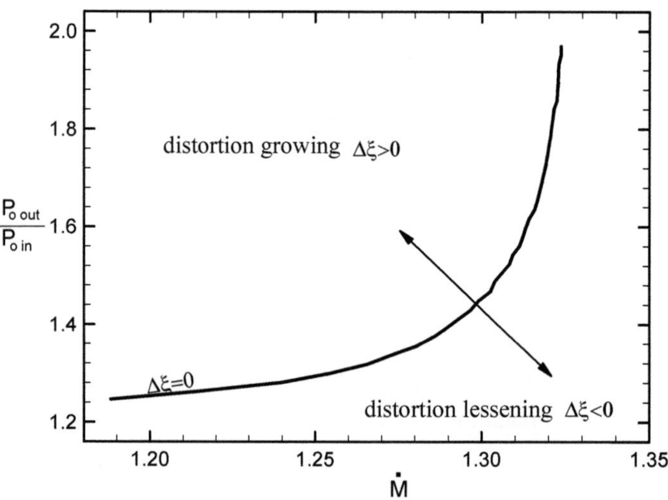

Fig. 1.15. The compressor critical performance with $\zeta(0) = 0.5$

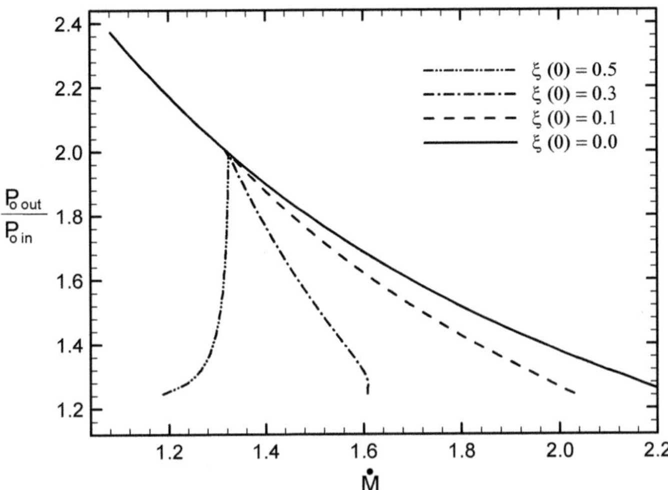

Fig. 1.16. The compressor critical performance with difference inlet distortion regions

The trend of ψ_p versus ϕ is similar to the compressor characteristic with uniform flow. Therefore, the curve of ψ_p versus ϕ in Fig. 1.17 is another way of describing the critical distortion line, termed the compressor critical characteristic because each point on it is corresponding to one point on the critical distortion line. On other hand, because the size of the distortion

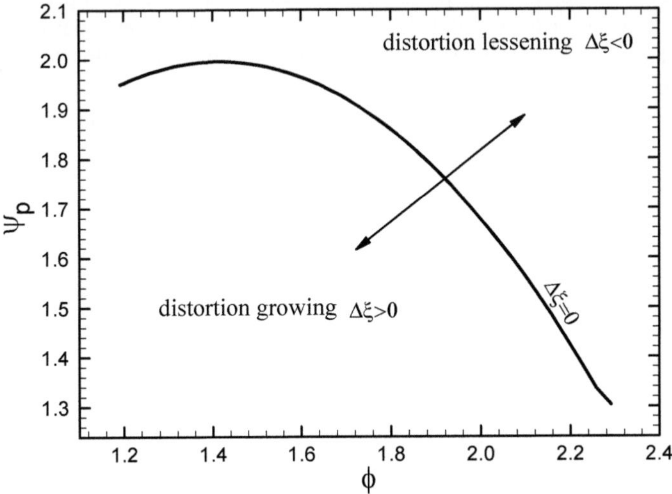

Fig. 1.17. The compressor critical characteristic

region has no effect on the compressor critical characteristic, there is an uniform compressor characteristic for compressor with and without inlet distortion. This phenomenon confirms the active role of compressor in determining the velocity distribution when compressor responds to an inlet flow distortion. In conclusion, the integral method is a feasible approach to produce qualitatively correct results for compressor, and it is hoping that to associate the current integral method with existing model, we could predict the onset of stall accurately.

1.5 Concluding Remarks

The integral method and previous results are investigated and developed. The results show that the distortion region may vanish downstream with the lessening of ξ when $d\alpha/dx > 0$. The results of the effects of inlet distorted parameters indicate that a low flux of mass, or even a back flow may be caused by the smaller inlet distorted velocity or larger flow angle. Excepting the ratio of drag-to-lift coefficients of the blade and the angle of flow as suggested by [8], the value of distorted inlet velocity is confirmed to be another essential parameter to control the distortion propagation. Among all parameters to control the distortion propagation, the angle of flow and the value of distorted inlet velocity are inlet parameters.

To express the distortion propagation correctly, a critical distortion line is proposed to describe the effect of two main inlet parameters: the angle of flow and the distorted inlet velocity, on the propagation of distortion. Further more, corresponding to the critical distortion line, the compressor performance and characteristic with distortion propagation are investigated.

This study is useful in understanding the axial physical behavior of compressor and the response of compressor to an inlet distortion, so as to construct a parameter which may correlate the inception of inlet distortion propagation. Finally, an improved model to predict the onset of compressor flow instability due to inlet distortion could be developed by means of the current integral method and results.

References

[1]   Christensen, D., Cantin, P., Gutz, D., Szucs, P.N., Wadia, A.R., Armor, J., Dhingra, M., Neumeier, Y. and Prasad, J.V.R., 2006, Development and demonstration of a stability management system for gas turbine engines. *ASME Turbo Expo,* GT2006-90324, Barcelona, Spain, May 8-11, 2006.

[2]   Chue, r. Hynes, T.P., Greitzer, E.M., Tan, C.S., and Longley, J.P., 1989, Calculations of inlet distortion induced compressor flow field instability. *International Journal of Heat and Fluid Flow,* **10(3)**: 211-223.

[3]   Day, I.J. , 1993, Active suppression of rotating stall and surge in axial compressors. *ASME Journal of Turbomachinery,* **115**: 40-47.

[4]   Dhingra, M., Neumeier, Y., Prasad, J.V.R., Breeze-Stringfellow, A., Shin, H-W. and Szucs, P.N., 2006, A stochastic model for a compressor stability measure. *ASME Turbo Expo*, GT2006-91182, Barcelona, Spain, May 8-11, 2006.

[5]   Greitzer, E.M. , 1980, Review-axial compressor stall phenomena. *ASME Journal of Fluids Engineering*, **102**: 134-151.

[6]   Greitzer, E.M. and Griswold, H.R., 1976, Compressor-diffuser interaction with circumferential flow distortion. *Journal of Mechanical Engineering Science*, **18(1)**: 25-38.

[7]   Harry, III, D.P. and Lubick, R.J., 1955, Inlet-air distortion effects on stall, surge, and acceleration margin of a turbojet engine equipped with variable compressor inlet guide vanes. *NACA Report*, NACA RM E54K26.

[8]   Kim, J.H., Marble, F.E., and Kim, C.–J., 1996, Distorted inlet flow propagation in axial compressors. In *Proceedings of the 6th International Symposium on Transport Phenomena and Dynamics of Rotating Machinery*, **2**: 123-130.

[9]   Liu Luxin, Neumeier, Y. and Prasad, J.V.R., 2005, Active flow control for enhanced compressor performance. AIAA paper-4017, *41st AIAA/ASME/ SAE/ASEE Joint Propulsion Conference & Exhibition*, Tucson Convention Center, Arizona, July 10-13, 2005.

[10]   Reid, C., 1969, The response of axial flow compressors to intake flow distortion. In *Proceeding of International Gas Turbine and Aeroengine Congress and Exhibition*, ASME Paper 69-GT-29.

[11]   Stenning, A.H., 1980, Inlet distortion effects in axial compressors. *ASME Journal of Fluids Eng.*, **102(3)**: 7-13.

Appendix 1.A   Fortran Program: Integral Method

This code solves the partial differential equation set of integral method (1.1), (1.2) and (1.3) by using the fourth-order Runge_Kutta method.

1.   The input data are

SITA1 : θ_0 flow angle at inlet, (1.6)

SKL   : k_L lift coefficient, (1.12)

DL   : k_D/k_L the ratio of drag-to-lift, (1.21)

Y(0)   : $\delta(0)$ vertical extension of distorted flow at inlet

DX   : Δx , spatial increment along x direction

XL   : L , ending position of x

ALFA(1,1): $\alpha(0)$, x- direction velocity increment in the distorted inlet region at inlet

ALFA(2,1): $\beta(0)$, y- direction velocity increment in the distorted inlet region

ALFA(3,1): $\alpha_0(0)$, x- direction velocity increment in the undistorted inlet region

ALFA(4,1): $\beta_0(0)$, y- direction velocity increment in the undistorted inlet region

ALFA(5,1): $P(0)$, pressure at inlet

2.   Assumptions:

Referential local flow directions relative to a stator and a rotor: SITAS=PSAIS=10 ($\theta^* = \psi^* = 10^0$).

At inlet, $\alpha_0(0) = \beta_0(0) = 1.0$, $\alpha(0) = \beta(0)$.

3.   Output results can be found in the data file: OUTPUT.DAT, in which, there are:

DY1   : $\delta(L) - \delta(0)$, the increment of vertical extension of distorted flow

DU0   : $\dfrac{\sqrt{\alpha_0{}^2 + \beta_0{}^2}}{2\gamma}$

DP12   :the pressure difference between outlet and inlet, (1.99)

DU   : $\dfrac{\sqrt{\alpha^2 + \beta^2}\,/\,\delta + \sqrt{\alpha_0{}^2 + \beta_0{}^2}\,/(1-\delta)}{2\gamma}$

DM   :no-dimensional mass flow rate, (1.94)

DP0   :the ratio of outlet to inlet total pressure, (1.97)

Some variables are explained in the code after "!".

```fortran
      PROGRAM RG_KT4
***********************************************************
*            COMPRESSOR   INTEGRAL   SOLVER            *
*            FUNCTIONS: VELOCITIES, PRESSURE           *
*            METHOD: FOURTH-ORDER RUNGE-KUTTA          *
***********************************************************
      PARAMETER (NDX=100000)
      DIMENSION  FXU(NDX),FX0U(NDX),FYU(NDX),FY0U(NDX),A(5)
      DIMENSION  X(NDX),Y(NDX),ALFA(5,NDX),RK(4,5,NDX),F(5,NDX)
      DOUBLE PRECISION RK,X,Y,ALFA,FXU,FX0U,FYU,FY0U,DX,F,
     &        SKD,SKL,SITA0,SITAS,PSAIS,GAMA,CK0,CK1,CK2,CK3,
     &        PI,DL,FX01,FX02,FX03,FX04,FX05,FX06,FX07,FX08,
     &        FY01,FY02,FX1,FX2,FX3,FX4,FX5,FX6,FX7,FX8,FY1,FY2,
     &        A,AL0,AL1,SI0,SI1,DSI0,DY,DY1,SITA1,DM,DM1,
     &        DP01,DP02,DP0,DP12,DU1,DU2,DU,DU0,A1,A2,A3
      OPEN(1,FILE='OUTPUT.DAT')
      PI=4.*ATAN(1.0)
******     PREPARE INITIAL DATA     *******
      SITA1=12.6
      SITAS=10.*PI/180.
      PSAIS=10.*PI/180.
      SKL=3.                              !     KL
      DL=1.0                              !     KD/KL
      SKD=DL*SKL                          !     KD
      Y(1)=0.0001                         !     DELTA/(PI*R)
      DY=0.1
      DSI0=0.2
      X(1)=0.                                  ! INITIAL POSITION OF X
      DX=0.001              ! SPACIAL INCREMENT ALONG X DIRECTION
      XL=1.                                    ! ENDING POSITION OF X
      ALFA(1,1)=0.06                      !     ALFA
```

```fortran
      ALFA(2,1)=ALFA(1,1)                      !     BETA
      ALFA(3,1)=1.                             !     ALFA0
      ALFA(4,1)=ALFA(3,1)                      !     BETA0
      ALFA(5,1)=0.0                            !     PRESSURE
1     SI0=SITA1
      SITA0=SITA1*PI/180.
      GAMA=TAN(SITA0)
      AL0=ALFA(1,1)
      DAL0=0.001
      AL1=ALFA(1,1)

10    CK1=Y(1)*ALFA(1,1)                               !     K1
      CK0=CK1+(1-CK1/ALFA(1,1))*ALFA(3,1)             !     K0
******     END OF INPUT    ********

******     ITERATIVE CALCULATION FROM X=0 TO X=500 **********
      I=1
20    I=I+1
          X(I)=X(I-1)+DX
******     FORCE COMPONENTS     ********
******     (1) DISTORTED REGION:
          FX1=ALFA(1,I-1)**2+((2.-ALFA(2,I-1))*GAMA)**2
          FX2=(2.-ALFA(2,I-1))*GAMA/ALFA(1,I-1)-TAN(PSAIS)
          FX3=(2.-ALFA(2,I-1))*GAMA/ALFA(1,I-1)*(1.-DL)+
     &          DL*TAN(PSAIS)
          FX4=ALFA(1,I-1)/(SQRT(ALFA(1,I-1)**2+
     &          ((2.-ALFA(2,I-1))*GAMA)**2))
          FX5=ALFA(1,I-1)**2+(ALFA(2,I-1)*GAMA)**2
          FX6=ALFA(2,I-1)*GAMA/ALFA(1,I-1)-TAN(SITAS)
          FX7=ALFA(2,I-1)*GAMA/ALFA(1,I-1)*(1.-DL)+
     &          DL*TAN(PSAIS)
          FX8=ALFA(1,I-1)/(SQRT(ALFA(1,I-1)**2+
     &          (ALFA(2,I-1)*GAMA)**2))
```

```fortran
        FXU(I)=SKL/4.*(FX1*FX2*FX3*FX4+FX5*FX6*FX7*FX8)
                                                        ! FX/U0**2
        FY1=1.+FX2*DL*(2.-ALFA(2,I-1))*GAMA/ALFA(1,I-1)
        FY2=1.+FX6*DL*ALFA(2,I-1)*GAMA/ALFA(1,I-1)
        FYU(I)=SKL/4.*(FX1*FX2*FY1*FX4-FX5*FX6*FY2*FX8)
                                                        ! FY/U0**2
******  (2) UNDISTORTED REGION:
        FX01=ALFA(3,I-1)**2+((2.-ALFA(4,I-1))*GAMA)**2
        FX02=(2.-ALFA(4,I-1))*GAMA/ALFA(3,I-1)-TAN(PSAIS)
        FX03=(2.-ALFA(4,I-1))*GAMA/ALFA(3,I-1)*(1.-DL)
     &          +DL*TAN(PSAIS)
        FX04=ALFA(3,I-1)/SQRT(ALFA(3,I-1)**2+
     &            ((2.-ALFA(4,I-1))*GAMA)**2)
        FX05=ALFA(3,I-1)**2+(ALFA(4,I-1)*GAMA)**2
        FX06=ALFA(4,I-1)*GAMA/ALFA(3,I-1)-TAN(SITAS)
        FX07=ALFA(4,I-1)*GAMA/ALFA(3,I-1)*(1.-DL)+DL*TAN(PSAIS)
        FX08=ALFA(3,I-1)/SQRT(ALFA(3,I-1)**2+
     &            (ALFA(4,I-1)*GAMA)**2)
        FX0U(I)=SKL/4.*                        ! FX,0/U0**2
     &        (FX01*FX02*FX03*FX04+FX05*FX06*FX07*FX08)
        FY01=1.+FX02*DL*(2.-ALFA(4,I-1))*GAMA/ALFA(3,I-1)
        FY02=1.+FX06*DL*ALFA(4,I-1)*GAMA/ALFA(3,I-1)
        FY0U(I)=SKL/4.*                        ! FY,0/U0**2
     &        (FX01*FX02*FY01*FX04-FX05*FX06*FY02*FX08)
******  END OF FORCE CALCULATION ********
***************************************************************
        CK2=CK1*(CK1-CK0)/(ALFA(1,I-1)-CK1)**2           ! K2
        CK3=1.+2.*CK1*(CK1-CK0)**2/(ALFA(1,I-1)-CK1)**3  ! K3
******  ITERATIVE CONSTANTS      ******
        DO 100 K=1,5
100     A(K)=ALFA(K,I-1)
        F(1,I)=(FXU(I)-FX0U(I))/(CK3*A(1))
        F(2,I)=FYU(I)/(GAMA*A(1))
        F(3,I)=CK2*F(1,I)
```

```fortran
      F(4,I)=FY0U(I)/(GAMA*A(3))-A(4)/A(3)*CK2*F(1,I)
      F(5,I)=(FXU(I)-F(1,I))/(GAMA**2*2)
      DO 101 K=1,5
101   RK(1,K,I)=DX*F(K,I)                ! K11,K12,K13,K14,K15
      DO 200 K=1,5
200   A(K)=ALFA(K,I-1)+RK(1,K,I)/2.
      F(1,I)=(FXU(I)-FX0U(I))/(CK3*A(1))
      F(2,I)=FYU(I)/(GAMA*A(1))
      F(3,I)=CK2*F(1,I)
      F(4,I)=FY0U(I)/(GAMA*A(3))-A(4)/A(3)*CK2*F(1,I)
      F(5,I)=(FXU(I)-F(1,I))/(GAMA**2*2)
      DO 201 K=1,5
201   RK(2,K,I)=DX*F(K,I)                ! K21,K22,K23,K24,K25
      DO 300 K=1,5
300   A(K)=ALFA(K,I-1)+RK(2,K,I)/2.
      F(1,I)=(FXU(I)-FX0U(I))/(CK3*A(1))
      F(2,I)=FYU(I)/(GAMA*A(1))
      F(3,I)=CK2*F(1,I)
      F(4,I)=FY0U(I)/(GAMA*A(3))-A(4)/A(3)*CK2*F(1,I)
      F(5,I)=(FXU(I)-F(1,I))/(GAMA**2*2)
      DO 301 K=1,5
301   RK(3,K,I)=DX*F(K,I)                ! K31,K32,K33,K34,K35
      DO 400 K=1,5
400   A(K)=ALFA(K,I-1)+RK(3,K,I)
      F(1,I)=(FXU(I)-FX0U(I))/(CK3*A(1))
      F(2,I)=FYU(I)/(GAMA*A(1))
      F(3,I)=CK2*F(1,I)
      F(4,I)=FY0U(I)/(GAMA*A(3))-A(4)/A(3)*CK2*F(1,I)
      F(5,I)=(FXU(I)-F(1,I))/(GAMA**2*2)
      DO 401 K=1,5
401   RK(4,K,I)=DX*F(K,I)                ! K41,K42,K43,K44,K45
      DO 105 K=1,5
          ALFA(K,I)=ALFA(K,I-1)+
     &      (RK(1,K,I)+2.*(RK(2,K,I)+RK(3,K,I))+RK(4,K,I))/6.
```

```fortran
105       CONTINUE
          Y(I)=CK1/ALFA(1,I)
******         OUTPUT        **********
      IF (X(I).LT.XL) GOTO 20
      DY1=Y(I)-Y(1)
      IF (DY*DY1.LT.0.) THEN
      DP01=(ALFA(3,1)**2/GAMA**2+ALFA(4,1)**2)/4+ALFA(5,1)
      DP02=(ALFA(3,I)**2/GAMA**2+ALFA(4,I)**2)/4+ALFA(5,I)
      DP0=DP02/DP01
      DM=((ALFA(1,I)-ALFA(3,I))*Y(I)+1.0)/(GAMA*2)
      DU1=SQRT(ALFA(3,I)**2+(ALFA(4,I)*GAMA)**2)*(1.-Y(I))
      DU2=SQRT(ALFA(1,I)**2+(ALFA(2,I)*GAMA)**2)*Y(I)
      DU=(DU1+DU2)/(GAMA*2)
      DU0=SQRT(ALFA(3,I)**2+(ALFA(4,I)*GAMA)**2)/(GAMA*2)
      DP12=ALFA(5,I)
      A1=SQRT(ALFA(1,I)**2+(ALFA(2,I)*GAMA)**2)/(GAMA*2)
      WRITE(*,*)ALFA(1,1),SITA1,DY1
      WRITE(1,210)DU0,DP12,DU,DM,DP0,A1,ALFA(1,1),SITA1,DY1
          SI0=SI0+DSI0
          SITA1=SI0
          IF (SITA1.GT.25.) GOTO 220
          DY=0.1
          GOTO 1
      END IF
      AL1=ALFA(1,1)
      DY=Y(I)-Y(1)
109   IF (ALFA(1,1).LT.1.01) THEN
          AL0=AL0+DAL0
          ALFA(1,1)=AL0
          ALFA(2,1)=AL0
          GOTO 10
      END IF
210   FORMAT (9F12.5)
310   FORMAT (7F10.5)
```

```
      CLOSE(1)
220   STOP
      END
```

* *

Appendix 1.B   Fortran Program: Critical Distortion Line

The variables used in this code are same as Appendix 1.A.

```fortran
      PROGRAM RG_KT4_CDL
************************************************************************
* COMPRESSOR  INTEGRAL  SOLVER FOR CRITICAL DISTORTION LINE       *
*              METHOD: FOURTH-ORDER RUNGE-KUTTA                   *
************************************************************************
      PARAMETER (NDX=100000)
      DIMENSION  FXU(NDX),FX0U(NDX),FYU(NDX),FY0U(NDX),A(5)
      DIMENSION  X(NDX),Y(NDX),ALFA(5,NDX),RK(4,5,NDX),F(5,NDX)
      DOUBLE PRECISION RK,X,Y,ALFA,FXU,FX0U,FYU,FY0U,DX,F,
     &          SKD,SKL,SITA0,SITAS,PSAIS,GAMA,CK0,CK1,CK2,CK3,
     &          PI,DL,FX01,FX02,FX03,FX04,FX05,FX06,FX07,FX08,
     &          FY01,FY02,FX1,FX2,FX3,FX4,FX5,FX6,FX7,FX8,FY1,FY2,
     &          A,AL0,AL1,SI0,SI1,DSI0,DY,DY1,SITA1,DM,DM1,
     &          DP01,DP02,DP0,DP12,DU1,DU2,DU,DU0,A1,A2,A3
      OPEN(1,FILE='OUTPUT.DAT')
      PI=4.*ATAN(1.0)
******    PREPARE INITIAL DATA       ********
      SITA1=12.6
      SITAS=10.*PI/180.
      PSAIS=10.*PI/180.
      SKL=3.                          !    KL
      DL=1.0                          !    KD/KL
      SKD=DL*SKL                      !    KD
      Y(1)=0.0001                     !    DELTA/(PI*R)
      DY=0.1
      DSI0=0.2
      X(1)=0.                         !    INITIAL POSITION OF X
      DX=0.001                        !    SPACIAL INCREMENT
                                      !    ALONG X DIRECTION
      XL=10.                          !    ENDING POSITION OF X
```

```
       ALFA(1,1)=0.06                         !      ALFA
       ALFA(2,1)=ALFA(1,1)                     !      BETA
       ALFA(3,1)=1.                            !      ALFA0
       ALFA(4,1)=ALFA(3,1)                     !      BETA0
       ALFA(5,1)=0.0                           !      PRESSURE
1      SI0=SITA1
       SITA0=SITA1*PI/180.
       GAMA=TAN(SITA0)
       AL0=ALFA(1,1)
       DAL0=0.001
       AL1=ALFA(1,1)


10     CK1=Y(1)*ALFA(1,1)                               !      K1
       CK0=CK1+(1-CK1/ALFA(1,1))*ALFA(3,1)              !      K0
******      END OF INPUT   ********


******      ITERATIVE CALCULATION FROM X=0 TO X=500 **********
       I=1
20     I=I+1
          X(I)=X(I-1)+DX
******      FORCE COMPONENTS      ********
******      (1) DISTORTED REGION:
          FX1=ALFA(1,I-1)**2+((2.-ALFA(2,I-1))*GAMA)**2
          FX2=(2.-ALFA(2,I-1))*GAMA/ALFA(1,I-1)-TAN(PSAIS)
          FX3=(2.-ALFA(2,I-1))*GAMA/ALFA(1,I-1)*(1.-DL)+
     &           DL*TAN(PSAIS)
          FX4=ALFA(1,I-1)/(SQRT(ALFA(1,I-1)**2+
     &           ((2.-ALFA(2,I-1))*GAMA)**2))
          FX5=ALFA(1,I-1)**2+(ALFA(2,I-1)*GAMA)**2
          FX6=ALFA(2,I-1)*GAMA/ALFA(1,I-1)-TAN(SITAS)
          FX7=ALFA(2,I-1)*GAMA/ALFA(1,I-1)*(1.-DL)+
     &           DL*TAN(PSAIS)
          FX8=ALFA(1,I-1)/(SQRT(ALFA(1,I-1)**2+
     &           (ALFA(2,I-1)*GAMA)**2))
```

```fortran
        FXU(I)=SKL/4.*(FX1*FX2*FX3*FX4+FX5*FX6*FX7*FX8)
                                             ! FX/U0**2
        FY1=1.+FX2*DL*(2.-ALFA(2,I-1))*GAMA/ALFA(1,I-1)
        FY2=1.+FX6*DL*ALFA(2,I-1)*GAMA/ALFA(1,I-1)

        FYU(I)=SKL/4.*(FX1*FX2*FY1*FX4-FX5*FX6*FY2*FX8)
                                             ! FY/U0**2
******  (2) UNDISTORTED REGION:
        FX01=ALFA(3,I-1)**2+((2.-ALFA(4,I-1))*GAMA)**2
        FX02=(2.-ALFA(4,I-1))*GAMA/ALFA(3,I-1)-TAN(PSAIS)
        FX03=(2.-ALFA(4,I-1))*GAMA/ALFA(3,I-1)*(1.-DL)
     &          +DL*TAN(PSAIS)
        FX04=ALFA(3,I-1)/SQRT(ALFA(3,I-1)**2+
     &              ((2.-ALFA(4,I-1))*GAMA)**2)
        FX05=ALFA(3,I-1)**2+(ALFA(4,I-1)*GAMA)**2
        FX06=ALFA(4,I-1)*GAMA/ALFA(3,I-1)-TAN(SITAS)
        FX07=ALFA(4,I-1)*GAMA/ALFA(3,I-1)*(1.-DL)+DL*TAN(PSAIS)
        FX08=ALFA(3,I-1)/SQRT(ALFA(3,I-1)**2+
     &              (ALFA(4,I-1)*GAMA)**2)
        FX0U(I)=SKL/4.*                       ! FX,0/U0**2
     &      (FX01*FX02*FX03*FX04+FX05*FX06*FX07*FX08)
        FY01=1.+FX02*DL*(2.-ALFA(4,I-1))*GAMA/ALFA(3,I-1)
        FY02=1.+FX06*DL*ALFA(4,I-1)*GAMA/ALFA(3,I-1)
        FY0U(I)=SKL/4.*                       ! FY,0/U0**2
     &              (FX01*FX02*FY01*FX04-FX05*FX06*FY02*FX08)
******  END OF FORCE CALCULATION ********
*****************************************************************
        CK2=CK1*(CK1-CK0)/(ALFA(1,I-1)-CK1)**2        ! K2
        CK3=1.+2.*CK1*(CK1-CK0)**2/(ALFA(1,I-1)-CK1)**3  ! K3
******  ITERATIVE CONSTANTS       ******
        DO 100 K=1,5
100     A(K)=ALFA(K,I-1)
        F(1,I)=(FXU(I)-FX0U(I))/(CK3*A(1))
        F(2,I)=FYU(I)/(GAMA*A(1))
```

```
      F(3,I)=CK2*F(1,I)
      F(4,I)=FYOU(I)/(GAMA*A(3))-A(4)/A(3)*CK2*F(1,I)
      F(5,I)=(FXU(I)-F(1,I))/(GAMA**2*2)
      DO 101 K=1,5
101   RK(1,K,I)=DX*F(K,I)                !    K11,K12,K13,K14,K15
      DO 200 K=1,5
200   A(K)=ALFA(K,I-1)+RK(1,K,I)/2.
      F(1,I)=(FXU(I)-FX0U(I))/(CK3*A(1))
      F(2,I)=FYU(I)/(GAMA*A(1))
      F(3,I)=CK2*F(1,I)
      F(4,I)=FYOU(I)/(GAMA*A(3))-A(4)/A(3)*CK2*F(1,I)
      F(5,I)=(FXU(I)-F(1,I))/(GAMA**2*2)
      DO 201 K=1,5
201   RK(2,K,I)=DX*F(K,I)                !    K21,K22,K23,K24,K25
      DO 300 K=1,5
300   A(K)=ALFA(K,I-1)+RK(2,K,I)/2.
      F(1,I)=(FXU(I)-FX0U(I))/(CK3*A(1))
      F(2,I)=FYU(I)/(GAMA*A(1))
      F(3,I)=CK2*F(1,I)
      F(4,I)=FYOU(I)/(GAMA*A(3))-A(4)/A(3)*CK2*F(1,I)
      F(5,I)=(FXU(I)-F(1,I))/(GAMA**2*2)
      DO 301 K=1,5
301   RK(3,K,I)=DX*F(K,I)                !    K31,K32,K33,K34,K35
      DO 400 K=1,5
400   A(K)=ALFA(K,I-1)+RK(3,K,I)
      F(1,I)=(FXU(I)-FX0U(I))/(CK3*A(1))
      F(2,I)=FYU(I)/(GAMA*A(1))
      F(3,I)=CK2*F(1,I)
      F(4,I)=FYOU(I)/(GAMA*A(3))-A(4)/A(3)*CK2*F(1,I)
      F(5,I)=(FXU(I)-F(1,I))/(GAMA**2*2)
      DO 401 K=1,5
401   RK(4,K,I)=DX*F(K,I)                !    K41,K42,K43,K44,K45

      DO 105 K=1,5
```

```fortran
          ALFA(K,I)=ALFA(K,I-1)+
     &        (RK(1,K,I)+2.*(RK(2,K,I)+RK(3,K,I))+RK(4,K,I))/6.
105       CONTINUE

          Y(I)=CK1/ALFA(1,I)


******          OUTPUT          **********
      IF (X(I).LT.XL) GOTO 20
      DY1=Y(I)-Y(1)
      IF (DY*DY1.LT.0.) THEN
      DP01=(ALFA(3,1)**2/GAMA**2+ALFA(4,1)**2)/4+ALFA(5,1)
      DP02=(ALFA(3,I)**2/GAMA**2+ALFA(4,I)**2)/4+ALFA(5,I)
      DP0=DP02/DP01
      DM=((ALFA(1,I)-ALFA(3,I))*Y(I)+1.0)/(GAMA*2)
      DU1=SQRT(ALFA(3,I)**2+(ALFA(4,I)*GAMA)**2)*(1.-Y(I))
      DU2=SQRT(ALFA(1,I)**2+(ALFA(2,I)*GAMA)**2)*Y(I)
      DU=(DU1+DU2)/(GAMA*2)
      DU0=SQRT(ALFA(3,I)**2+(ALFA(4,I)*GAMA)**2)/(GAMA*2)
      DP12=ALFA(5,I)
      A1=SQRT(ALFA(1,I)**2+(ALFA(2,I)*GAMA)**2)/(GAMA*2)
      WRITE(*,*)ALFA(1,1),SITA1,DY1
      WRITE(1,210)ALFA(1,1),SITA1,DY1,DU0,DP12,DU,DM,DP0,A1
          SI0=SI0+DSI0
          SITA1=SI0
          IF (SITA1.GT.25.) GOTO 220
          DY=0.1
          GOTO 1
      END IF
      AL1=ALFA(1,1)
      DY=Y(I)-Y(1)

109   IF (ALFA(1,1).LT.1.0) THEN
          AL0=AL0+DAL0
          ALFA(1,1)=AL0
```

```
        ALFA(2,1)=AL0
        GOTO 10
    END IF
210 FORMAT (9F12.5)
    CLOSE(1)
220 STOP
    END
```

****

Chapter 2
Stall Prediction of In-flight Compressor due to Flaming of Refueling Leakage near Inlet

In this chapter, the critical distortion line and the integral method explained in Chap. 1 are extended further to investigate more practical applications about the propagation of strong distortion at inlet of an axial compressor. The practical applications, such as the inlet conditions of flaming of leakage fuel during mid-air refueling process, are implemented to show the details of the numerical methodology used in analysis of the axial flow compressor behavior and the propagation of inlet distortion. From the viewpoint of compressor efficiency, the propagation of inlet flow distortion is further described by compressor critical performance and its critical characteristic. The simulated results present a useful physical insight to the significant effects of inlet parameters on the distortion extension, velocity, and compressor characteristics. The distortion level, flow angle and the size of distortion area at compressor inlet, and the rotor blade speed are found being the major parameters affecting the mass flow rate of engine.

2.1 Introduction

The inlet flow distortion may cause the rotating stall, surge, or a combination of both. An inlet distortion is often encountered when the real flow is associated with some degree of angle of attack to the engine nacelle. Take-off, landing and gust encounter are some typical examples of such situation. On the other hand, the inlet distortion can also occur when the leakage of fuel enters the compressor and consumes part of the mass during the air to air refueling.

Since a severe inlet distortion may lead to a stall of fan blade and fatal loss of engine power, the evaluation of inlet distortion effects is very important in the engine safety problem. If such evaluation can be measured quantitatively, the designer may be able to design the compressor stage with a minimum effect of inlet distortion ([3], [4] and [5]).

With the rapid development in computational sciences, the numerical simulation of complete three-dimensional flows within multiple stages of compressor is becoming more effective and practical in design application. Nevertheless, many CFD codes have to be converted for parallel computation in recent year when it is implemented in large-scale simulations ([2], [8] and [11]), because such a complex simulation still require huge computing resources far exceeding the practical limits of most single-processor supercomputers. To predict the distorted performance and distortion attenuation of an axial compressor without using CFD codes, it is essential to make significant simplifications and therefore, some elegance and detail of description must be sacrificed.

By using a simple integral method Kim *et al.* [6] successfully calculated the qualitative trend of distorted performance and distortion attenuation of an axial compressor. The integral method is further modified [7] to simulate and analyze the effects of the parameters of inlet distortions on the trend of downstream flow feature in compressor. Because the distorted velocity and incident angle in inlet are the two essential inlet parameters to control the distortion in propagation, Ng *et al.* [7] proposed a critical distortion line, which include the combining effects of both inlet parameters. With introduction of the critical distortion line, the downstream flow status in compressor can be determined concurrently. This chapter illustrates a practical example on how to apply the proposed critical distortion line and integral method to analyze the axial flow compressor behavior and hence the propagation of inlet distortion.

Kim *et al.* [6] has concluded that the two most important parameters to control the distortion propagation are the drag-to-lift ratio of the blade and the inlet flow angle. Taguchi method ([9] and [10]) is used here to verify Kim *et al*'s findings on the parameters influencing the flow through the compressor.

2.2 Inlet Flow Condition

As an example, consider a situation of air-to-air refueling when a leakage of fuel enters the engine nacelle together with the inlet air. Some of the smaller droplets of leakage fuel vaporize and form a vapor-air mixture. When this mixture reach the fan blade or first stage of compressor with temperature of about $15\,^{\circ}C$ (much higher value if past the IGV) during refueling, the mixture may well be within the range of flammability, especially for the more volatile wide-cut fuel. The upper flammability temperature limit depends on the vapor pressure of the fuel [1].

The vapor-air mixture ignites itself when entering and leaving the guide vanes, and consumes part of mass flux entering the first stage of compressor. The inlet flow to the compressor is thus distorted. This inlet flow condition can be simplified into two regions: undistorted region with a normal mass flux and distorted region with an inadequate mass flux due to the flaming of leakage fuel. Each region is assumed to have a uniform velocity distribution respectively, and a same inlet flow angle in both regions with the guide vanes in-place.

The dimensionless velocity components in each of the regions are defined as:

distorted region:
$$u = \alpha U_0 \tag{2.1a}$$

$$v = \beta V_0 \tag{2.1b}$$

undistorted region:
$$u_0 = \alpha_0 U_0 \tag{2.1c}$$

$$v_0 = \beta_0 V_0 \tag{2.1d}$$

The inlet velocity has an flow angle of θ_0, and,

$$\overline{\gamma} = \tan\theta_0 = V_0/U_0 \tag{2.2}$$

where U_0 and V_0 are the x- and y- components of reference inlet velocity respectively. The distorted velocity coefficients α and β are the velocity fractions of the referenced inlet velocity in the distorted inlet region, and the undistorted velocity coefficients α_0 and β_0 are the velocity fractions of the referenced inlet velocity in the undistorted inlet region respectively. u and v are the x- and y- components of distorted velocity, and u_0 and v_0 are the x- and y- components of undistorted velocity respectively. To ease in computation, we assume $\alpha_0(0) = \beta_0(0)$, and $\alpha(0) = \beta(0)$.

2.3 Computational Domain

The computational domain considered here is a two-dimensional distorted inviscid flow through an axial compressor as schematized in Fig. 2.1. The distorted inlet

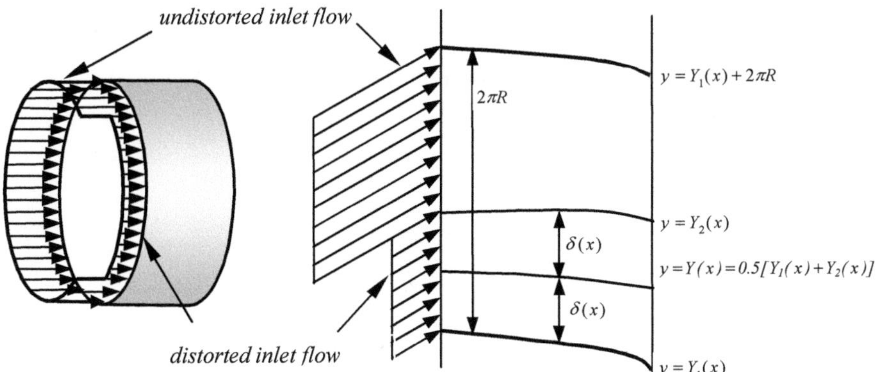

Fig. 2.1. Distorted inlet flow and its two-dimensional schematic

flow is simplified as a uniform mass loss in a specified distorted region. If the circumferential range of distorted inlet flow is assumed as 2δ, the circumferential extension of undistorted flow would then be $2(\pi R - \delta)$, and the relative circumferential size of distorted flow is $\delta/(\pi R)$.

Here, we assume that the distorted flow occupied half area of the cross section in the inlet region of compressor, i.e., $\xi(0) = \delta/(\pi R) = 0.5$. In fact, from the critical distortion lines as shown in Fig. 2.2, the relative size of distorted region at inlet has no effect on the propagation of inlet distortion.

For convenience in illustrating the effects of inlet distortion level and inlet flow angle on the propagation of inlet distortion, the critical distortion line is redrawn on the plane of coordinates ($\Gamma(0)$, θ_0) as shown by Fig. 2.3.

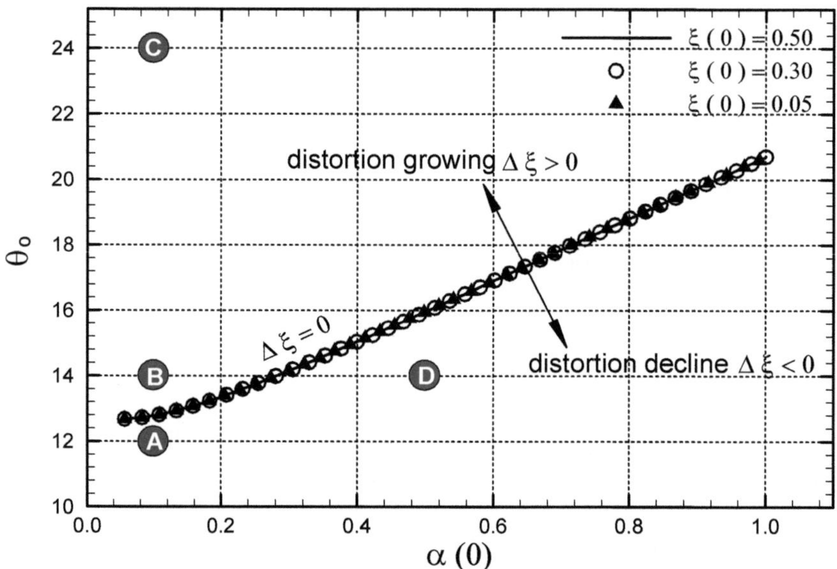

Fig. 2.2. Critical distortion line with different inlet size of distortion region

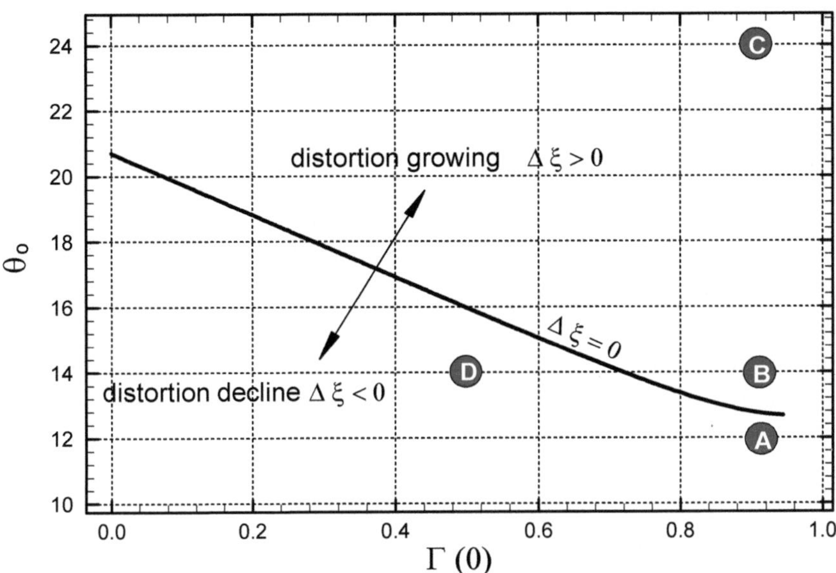

Fig. 2.3 Critical distortion line using inlet distortion level

2.4 Application of Critical Distortion Line

Consider a very extreme situation where most of the inlet mass, say, *90%*, is burned in the distorted region in which the inlet velocity coefficient of undistorted flow is set as unity ($\alpha_0(0)=1.0$) whereas the inlet velocity coefficient of distorted flow is *0.1* ($\alpha(0)=0.1$).

The loss of mass flow in inlet can be represented using distortion level:

$$\Gamma(x) = 1 - \frac{\alpha(x)}{\alpha_0(x)} \tag{2.3}$$

The smaller $\alpha(0)$ thus means higher inlet distortion level or a severe loss of mass flow rate. It is obvious that the definition of distortion level in representing the loss of mass flow is more intuitive than using the distorted velocity coefficient. For the case with $\alpha_0(0)=1.0$ and $\alpha(0)=0.1$, the distortion level at inlet is $\Gamma(0)=0.9$. This is a severe distortion case with a high initial distortion level.

At a very small inlet flow angle, such as, $\theta_0 = 12°$, in the coordinates plane of $\alpha(0)$ and θ_0 as illustrated in Fig. 2.2, the corresponding point of this case, $\alpha(0)=0.1$ and $\theta_0 = 12°$, can be determined. This point, named point A (or case A) here, is located below the critical distortion line. The critical distortion line divides the plane of coordinates $\alpha(0)$ and θ_0 into two areas [7]. In the cases pointed above the line, the propagation of inlet distortion will grow at downstream; on the contrary, in the cases pointed below the line, the distortion of inlet will decline along the axis of compressor. Apparently, the point A indicates the decline feature of given case. Next, by increasing the inlet flow angle to a higher value, say $\theta_0 = 14°$, the case pointed on the plane of $\alpha(0)$ and θ_0 by B is an unstable example because the point B (case B) is located above the critical line, the inlet distortion thus will grow in downstream direction of compressor. If the inlet flow angle is further increased to $\theta_0 = 24°$ (point C or case C), the inlet distortion will grow faster than that of case B at downstream of compressor since point C is further from the critical distortion line than point B.

However, if there is less air being burnt in the distorted region, the limit of stable inlet flow angle will increase with the higher value of $\alpha(0)$. In the other words, the cases with same inlet flow angle but larger inlet distorted velocity coefficient (or lower inlet distortion level) are more stable. For an example, a case with $\alpha(0)=0.5$ and inlet flow angle of $14°$ ($\theta_0 = 14°$), as pointed by D in Fig. 2.2 and Fig. 2.3, has the same inlet flow angle with point B, but a lower inlet distortion level of $\Gamma(0)=0.5$. Unlike case B, case D is a stable situation and its propagation of inlet distortion will decline in the axial direction of compressor.

2.5 Application of Integral Method

The integral method (Ng *et al.* (2002)) provided a set of integral equation:

$$\alpha \frac{d\alpha}{dx} = \frac{1}{K_3}\left(\frac{F_x - F_{x,0}}{U_0^2}\right) \tag{2.4a}$$

$$\alpha \frac{d\beta}{dx} = \frac{1}{\bar{\gamma}}\left(\frac{F_y}{U_0^2}\right) \tag{2.4b}$$

$$\frac{d\alpha_0}{dx} = K_2 \frac{d\alpha}{dx} \tag{2.4c}$$

$$\alpha_0 \frac{d\beta_0}{dx} = \frac{1}{\gamma}(\frac{F_{y,0}}{U_0^2}) \tag{2.4d}$$

$$\frac{d}{dx}(\frac{p}{\rho}) = U_0^2(\frac{F_x}{U_0^2} - \alpha \frac{d\alpha}{dx}) \tag{2.4e}$$

The integral equations can be solved numerically using the 4th-order Runge-Kutta method. By solving the equations, five variables are obtainable. They are two distorted velocity coefficients α and β, two undistorted velocity coefficients α_0 and β_0, and one static pressure (p/ρ).

For cases with severe distortion due to much of the air being burned, such as the cases A, B, and C, the inlet relative distorted area is $\xi(0) = \delta/(\pi R) = 0.5$, and the inlet distortion level $\Gamma(0) = 0.9$ and the relative velocity in the distorted region is $\frac{\alpha(0)}{\alpha_0(0)} = 0.1$. Using the critical line, one can determine that the case A is a stable condition whereas the cases B and C are the unstable situations.

Further more, the relative change rate of the parameter f is defined as:

$$\varepsilon(f) = \frac{f(x) - f(0)}{f(0)} \% \tag{2.5}$$

where $f(0)$ is the basal value and $f(0)$ is the current value. This definition is used to calculate the relative change rate of inlet velocity coefficients, distortion level and the size of the distortion at inlet.

The propagation of distortion through a ten-stage compressor can then be simulated easily with the integral method proposed [7]. In case A, the distorted region size declines from a value of 0.5 at inlet to 0.49465 at outlet as shown in Fig. 2.4. The change rate of distorted region for case A is $\varepsilon(\xi) = -1.07\%$. Similarly, the change rate for case B can be obtained by $\varepsilon(\xi) = 1.554\%$, since the size of distorted region at outlet for case B is 0.50777. The propagation of distortion for cases A, B, C and D are shown in Fig. 2.4.

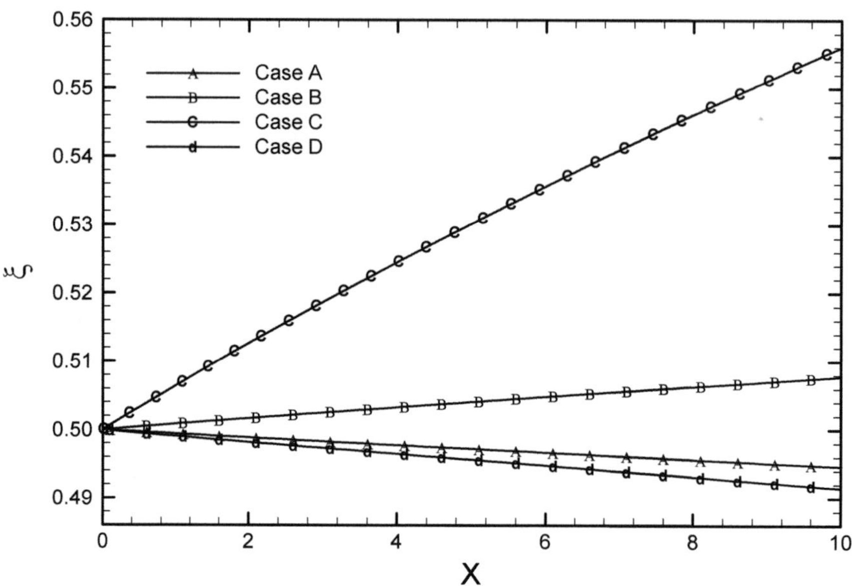

Fig. 2.4 The propagation of distorted flow for cases A, B, C and D

On the other hand, the change rates of velocity coefficients and distortion level along axial direction of compressor can also be predicted using integral method, and the simulated results for all the four cases are summarized in As indicated in Figs. 2.4, 2.5 and 2.6, and Table 2.1, case C is the worst situation in the four cases for the propagation of distortion. In case C, the size of distorted region at inlet is $\xi(0) = 0.5$, and grows up to $\xi(10) = 0.55605$ at outlet; its distortion level grows up to $\Gamma(10) = 0.92016$ from a value of $\Gamma(0) = 0.9$ at inlet. On the other hand, for case C, the relative velocity coefficient in distorted region decreases down to $\dfrac{\alpha(10)}{\alpha_0(10)} = 0.07984$ at outlet from a value of $\dfrac{\alpha(0)}{\alpha_0(0)} = 0.1$ at inlet (here the value of $\xi(x)$, $\Gamma(x)$, $\alpha(x)$, and $\alpha_0(x)$ when $x = 0$ and 10 are shown in Table 2.1). The sharp increases of distorted area and distortion level may cause large oscillations of mass flow rate or back flow, and the flow becomes unstable, with likelihood of surge phenomenon.

Table 2.1 Relative change rates of the velocity coefficients, distortion levels and the sizes of distorted region in the axial range of *[0,10]*

Cases	A	B	C	D
$\alpha(0)$	0.10000	0.10000	0.10000	0.50000
$\alpha(10)$	0.10108	0.09847	0.08992	0.50867
$\varepsilon(\alpha)$ (%)	1.080	-1.530	-10.080	1.734
$\alpha_0(0)$	1.00000	1.00000	1.00000	1.00000
$\alpha_0(10)$	0.98941	1.01579	1.12624	0.98323
$\varepsilon(\alpha_0)$ (%)	-1.059	1.579	12.624	-1.677
$\Gamma(0)$	0.90000	0.90000	0.90000	0.50000
$\Gamma(10)$	0.89784	0.90306	0.92016	0.48265
$\varepsilon(\Gamma)$ (%)	-0.240	0.340	2.240	-3.469
$\xi(0)$	0.50000	0.50000	0.50000	0.50000
$\xi(10)$	0.49465	0.50777	0.55605	0.49147
$\varepsilon(\xi)$ (%)	-1.070	1.554	11.210	-1.706

Note: for the calculation of $\varepsilon(f)=\dfrac{f(10)-f_0}{f_0}\%$, f_0 is the basal value at $x=0$.

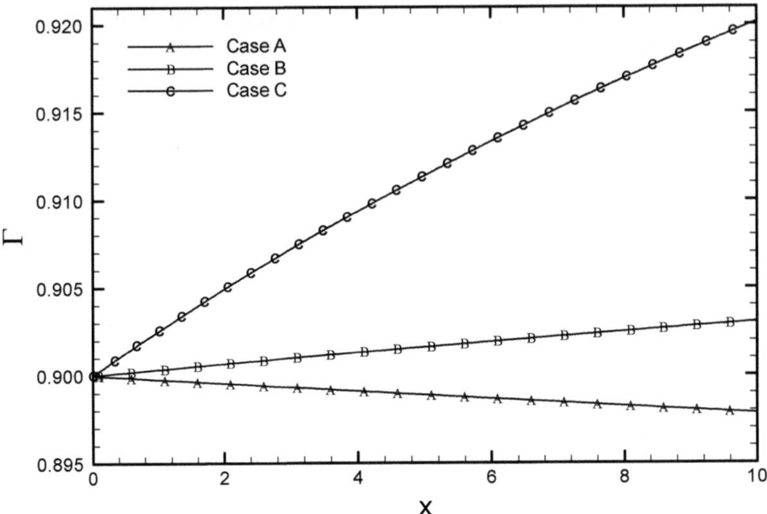

Fig. 2.5. The propagation of distortion level for cases A, B, and C

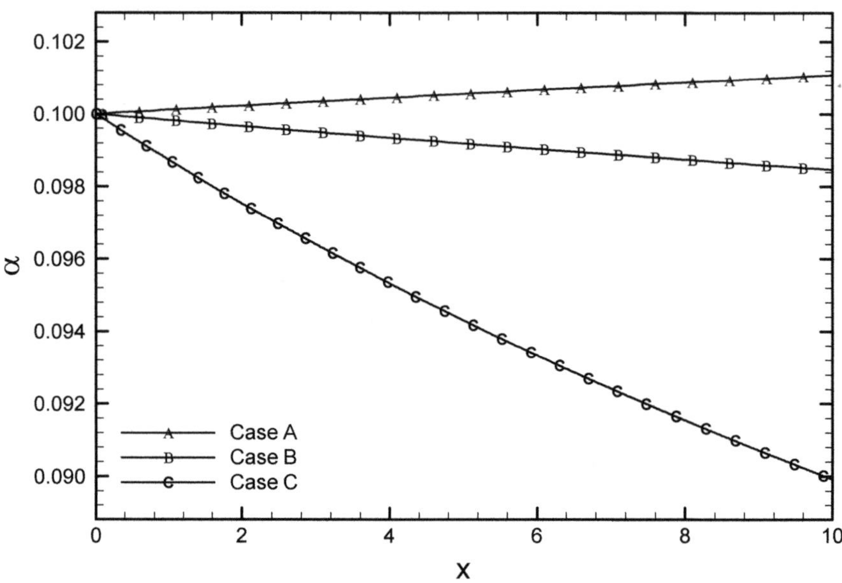

Fig. 2.6. The propagation of distorted velocity coefficients for cases A, B, and C

2.6 Compressor Characteristics

For investigating the compressor characteristics with inlet distortion, one could begin with case of significant inlet distortion, and decrease the distortion level gradually. The simulation is started with case C using conditions of: $\xi(0)=0.5$, $\theta=24°$, and $\Gamma(0)=0.9$ as a typical distortion example due to the obvious unstable condition of case C. Next, the distortion level is reduced by increasing the relative value between the velocities in distorted and undistorted regions at inlet. In the other words, $\alpha_0(0)=1.0$ is fixed and the distorted velocity at inlet, $\alpha(0)$, is increased from 0.1 to the undistorted value of unity, thus the distortion level is changed from $\Gamma(0)=0.9$ to $\Gamma(0)=0$. When $\alpha(0)=\alpha_0(0)=1.0$, or $\Gamma(0)=0$, the distortion disappears. A curve of compressor characteristic corresponding to different inlet distortion levels can be obtained as shown in Fig. 2.7.

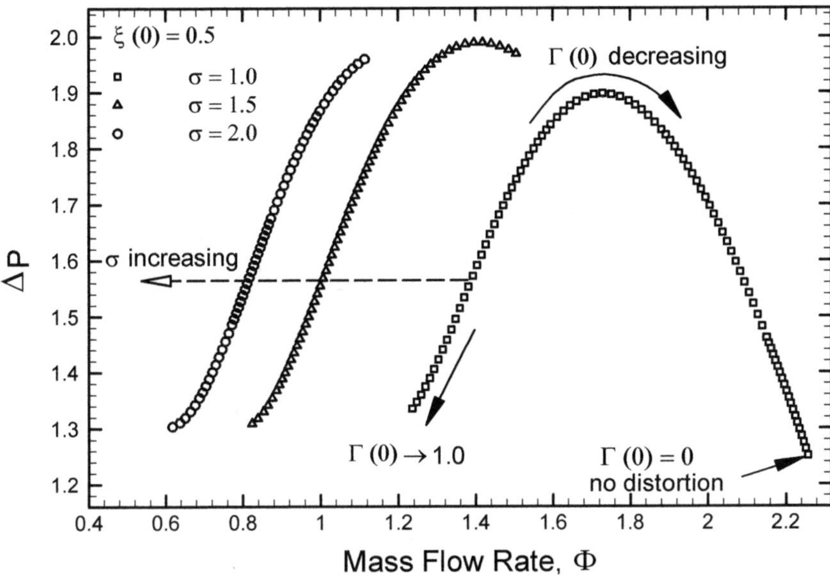

Fig. 2.7. Compressor characteristics with different rotor blade speeds for a constant flow angle at inlet, $\theta_0=24^o$

In Fig. 2.7, ΔP is the non-dimensional pressure difference between the inlet and outlet; Φ is non-dimensional mass flow rate:

$$\Delta P = \frac{p(10) - p(0)}{\rho U^2} \tag{2.6}$$

$$\Phi = \frac{(\alpha - \alpha_0)\xi + \alpha_0}{\gamma \sigma} \tag{2.7}$$

Here, $p(0)$ and $p(10)$ are static pressures at inlet and outlet, respectively; σ is the relative value of the rotor blade speed to circumferential velocity in the undistorted region at inlet:

$$\sigma = \frac{\omega R}{V_0} \tag{2.8}$$

2.6.1 Effects of Rotor Blade Speed (σ)

By changing the rotor blade speed σ, its effect on the compressor characteristic for a constant flow angle at inlet ($\theta_0 = 24°$) can be predicted and illustrated in Fig. 2.7.

In the situation of compressor with a stronger inlet distortion, the pressure difference throughout the compressor increases rapidly with the increase of mass flow rate. On the other hand, when there is lower distortion level at inlet, the pressure through the compressor may not increase, or even decrease for some cases with smaller rotor blade speed, such as in the case of $\sigma = 1.0$. The calculated results suggested that the peak of the curve for the case of $\sigma = 1.0$ occurred at around $\Gamma(0) = 0.46$ and $\alpha(0) = 0.54$.

As to the effects of σ, an increase in the value of σ will induce the decrease of relative velocity V_0 from the (2.8). Using the definition of γ :

$$\bar{\gamma} = tan\,\theta_0 = V_0 / U_0 \tag{2.9}$$

because $\bar{\gamma}$ is a fixed value in the current work, the value of U_0 will thus also decrease. Therefore the higher σ values produce lower mass flow rate with smaller velocity, U_0 as shown in Fig. 2.7.

For the case of fuel leakage as mentioned previously in the worse situation (such as the case C), there is a higher distortion level of $\Gamma(0) = 0.9$. In the other words, in the distorted region at inlet of compressor, there is only 10% of mass through flow rate into the inlet. The non-dimensional total mass flow rate decreases from 1.12302 to 0.61766 as compared with the case without distortion as indicated by the circle-symbols ($\sigma=2.0$) in Fig. 2.7. The relative change rate of non-dimensional total mass flow rate is (Table 2.2):

$$\varepsilon(\Phi) = \frac{0.61766 - 1.12302}{1.12302}\% = -45.0\% \tag{2.10}$$

Similarly, the relative change rate of pressure rise through the compressor due to distortion of $\Gamma(0) = 0.9$ is calculated by Table 2.2:

$$\varepsilon(\Delta P) = \frac{1.30386 - 1.96371}{1.96371}\% = -33.6\% \tag{2.11}$$

Table 2.2 Relative change rates of the mass flow rates and pressure rise when $\Gamma(0)$ is changed in the range of $[0.9, 0]$

$\xi(0)$	0.5	0.4	0.3	0.2	0.1
$\Phi\big|_{\Gamma(0)=0.9}$	0.61766	0.71873	0.81980	0.92087	1.02194
$\varepsilon(\Phi)$ (%)	-45.0	-36.0	-27.0	-18.0	-9.0
$\Delta P\big|_{\Gamma(0)=0.9}$	1.30386	1.30471	1.30595	1.30793	1.31186
$\varepsilon(\Delta P)$ (%)	-33.60	-33.56	-33.50	-33.39	-33.19

Note: when $\Gamma(0)=0$, the values $\Phi\big|_{\Gamma(0)=0}=1.12302$ and $\Delta P\big|_{\Gamma(0)=0}=1.96371$ are taken as the basal value for the calculation of ε. $\sigma=2.0$.

2.6.2 Effects of Inlet Distorted Region Size ($\xi(0)$)

The size of distorted region at inlet has no effect on the critical distortion line, but it has a significant effect on the compressor characteristic (Fig. 2.8). In all of the cases mentioned previously, the size of distorted region in inlet $\xi(0)$ is taken as 0.5. For investigating the effects of $\xi(0)$ on the compressor characteristic, the circle-symbols curve with $\xi(0)=0.5$ in Fig. 2.7 is used to compare with another cases with different $\xi(0)$, and shown in Fig. 2.8.

Each curve in Fig. 2.8 is obtained using the similar procedure as depicted in Fig. 2.7: fixing all parameters except $\Gamma(0)$, and decreasing $\Gamma(0)$ from 0.9 to 0. As expected, in the case with a higher level of distorted region, the total mass flow rate is smaller. Certainly, when there is zero distortion, $\Gamma(0)=0$, all curves intersect at a common point. However, the size of distorted region has no effect on the pressure rise.

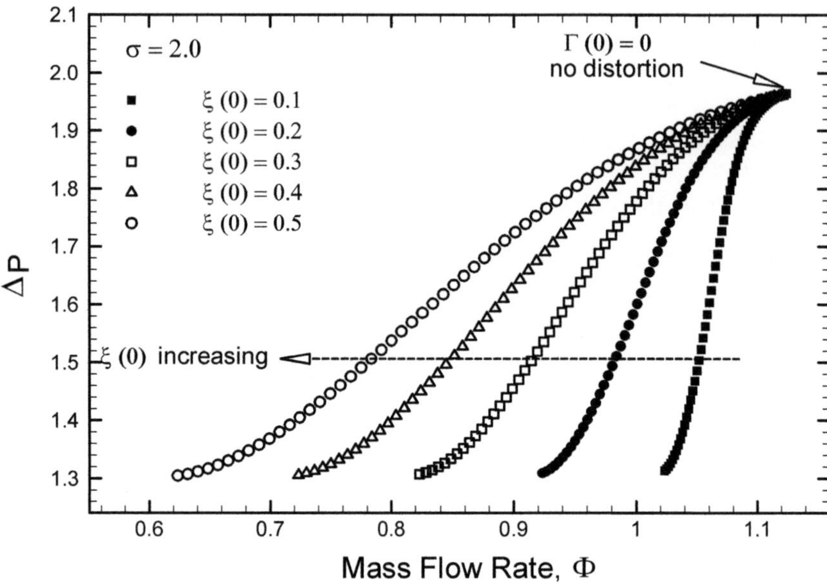

Fig. 2.8. Compressor characteristics with different inlet distortion sizes for a constant flow angle at inlet, $\theta_0=24^o$

In returning to the problem of fuel leakage, the value of $\varepsilon(\Delta P)$ is close to a constant of about *33%* for different $\xi(0)$, but $\varepsilon(\Phi)$ reduces linearly towards zero (from negative value) with the decreasing of $\xi(0)$ as shown in Fig. 2.9 and Table 2.2. In this table, when $\Gamma(0) = 0$, $\Phi\big|_{\Gamma(0)=0} = 1.12302$ and $\Delta P\big|_{\Gamma(0)=0} = 1.96371$ are the basal value of the mass flow rate and pressure rise, and $\Phi\big|_{\Gamma(0)=0.9}$ and $\Delta P\big|_{\Gamma(0)=0.9}$ are the current value of mass flow rate and pressure rise when $\Gamma(0) = 0.9$ with different inlet distorted region sizes at inlet.

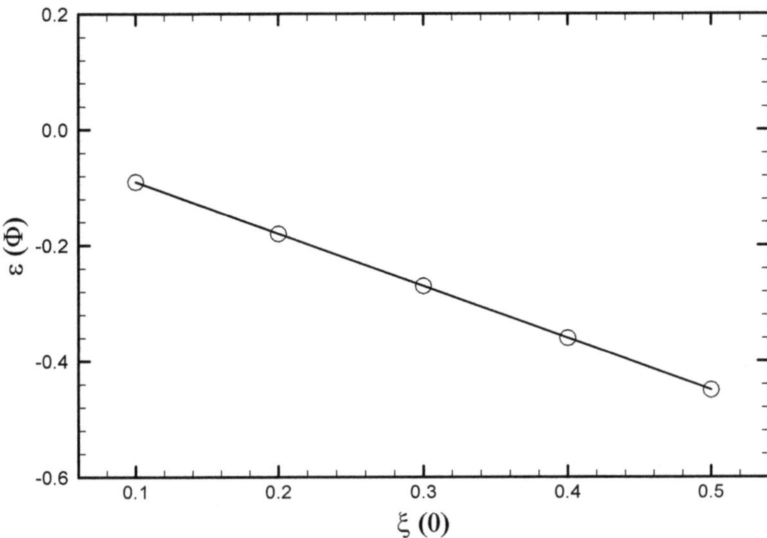

Fig. 2.9 The relative change rate of mass flow rate with different inlet distortion sizes, with $\sigma=2.0$, $\xi=0.5$ and $\theta_0=24^o$

2.7 Concluding Remarks

The current work predicted the propagation of inlet distortion in multistage compressor, and investigated the compressor characteristic with an inlet distortion by employing the integral method and critical distortion line.

As a typical case, a strong inlet distortion caused by fuel leakage is studied. When the distortion level arrives upon $\Gamma=90\%$, the distortion will grow rapidly along the axial direction of the compressor so long as the incident angle at inlet is higher than 12.8°. Normally, the incident angle at inlet is much higher than this critical value, and the higher the incident angle at inlet, the more the propagation of distortion would grow. Conclusively, the distortion with $\Gamma=90\%$ will induce the rapid propagation of distortion and produce an unstable flow in axial compressor. In this case, when the incident angle at inlet is taken as $\theta_0=24^\circ$, the size of distorted region will grow more than 11% in a ten-stage compressor.

In a typical case, when half of the inlet region is distorted, say $\xi(0)=0.5$, by comparing with the situation of no distortion, the non-dimensional mass flow rate will decrease more than 45% in a ten-stage compressor.

The propagation or/and redistribution of inlet flow distortion is affected by both the inlet distortion level and flow angle. Higher inlet distortion level or larger flow angle tends to induce the flow through the compressor unstable. On the other hand, the mass flow rate is affected by both the rotor blade speed and the size of distorted region. The higher relative rotor blade speed or larger distorted region size likely to produce a higher loss of the mass flow rate.

In all, the above situations of unstable flow in compressor or the loss of mass flow rate may cause the rotating stall or even surge or a combination of them, and should be avoided by understanding and controlling the effecting parameters. The current study indicates the major effecting parameters including the inlet distortion level, flow angle, distorted region size at inlet, and rotor blade speed.

Similar hypothetical study on the causes of stall and surge phenomena in compressor such as "bird strike" and large separation of boundary layer air from fuselage could be performed with the present approach.

References

[1]   Aviation fuels technical review (FTR-3), Chevron Products Company, A division of Chevron U.S.A. Inc. 2002.

[2]   Chen J.P. and Briley W.R., 2001, A parallel flow solver for unsteady multiple blade row turbomachinery simulations. ASME paper 2001-GT-0348.

[3]   Chue R., Hynes T.P., Greitzer E.M., Tan C.S. and Longley J.P., 1989, Calculations of inlet distortion induced compressor flow field instability. *International Journal of Heat and Fluid Flow*, **10(3)**: 211-223.

[4]   Day I.J., 1993, Active suppression of rotating stall and surge in axial compressors. *ASME Journal of Turbomachinery*, **115**: 40-47.

[5]   Greitzer E.M., 1980, Review: axial compressor stall phenomena. *ASME Journal of Fluids Engineering*, **102**: 134-151.

[6]   Kim J.H., Marble F.E. and Kim C–J., 1996, Distorted inlet flow propagation in axial compressors. In *Proceedings of the 6th International Symposium on Transport Phenomena and Dynamics of Rotating Machinery*, **2**: 123-130.

[7]   Ng E.Y-K., Liu N., Lim H.N. and Tan T.L., 2002, Study On The Distorted Inlet Flow Propagation In Axial Compressor Using An Integral Method, *Computational Mechanics*, **30(1)**: 1-11.

[8]   Stenning A.H., 1980, Inlet distortion effects in axial compressors. *ASME Journal of Fluids Eng.*, **102(3)**: 7-13.

[9]   Taguchi, G., 1993, Taguchi On Robust Technology Development: Bringing Quality Engineering Upstream, New York: ASME Press.

[10]   Taguchi, G., 1986, Introduction To Quality Engineering: Designing Quality Into Products And Processes, Japan: Asian Productivity Organization.

[11]   Wellborn S.R. and Delaney R.A., 2001, Redesign of a 12-stage axial-flow compressor using multistage CFD. ASME paper 2001-GT-0315.

Chapter 3
Parametric Study of Inlet Distortion Propagation in Compressor with Integral Approach and Taguchi Method

As mentioned and discussed in previous chapters, the integral method can successful to describe the qualitative trend of distorted inlet flow propagation in the axial compressors. Generally, integral method is applied to the problems of distorted inlet flow, and the relationships and the effects that some of the key parameters would have on the propagation of inlet distortion flow were predicted in qualitative trend, as illustrated in Chap. 1. In this chapter, a Taguchi's quality control method [12] will be adopted to justify the integral method and its research results. The results from Taguchi's quality control method indicate that the influence of major parameters on the inlet distortion propagation can be ranked as, the most one of the ratio of drag-to-lift coefficient, then the inlet distorted velocity coefficient, and the least one of inlet flow angle. This conclusion is different from that in Kim *et al.*'s research, reason being the later was carried out using only several cases with integral method. In comparison, when Taguchi quality control method is used, the prediction of degrees of influence by the parameters on the distortion propagation is more reasonable and accurate.

3.1 Introduction

The gas turbine engine has contributed greatly to the advancement of current flight capabilities in terms of aircraft performance and range. The propulsive power of the gas turbine has increased since World War II through higher cycle pressure ratios and turbine inlet temperature. The compressor is a key component to this evolution.

In general, it is more difficult to attain high efficiency on the compressor stages. Compressors must achieve high efficiency in blade rows in diffusing flow fields. However, stable operation of the engine depends on the range of stable operation of the compressor and the blade row stall characteristics determine the limit of stable operation.

Compressor performance is normally characterized by pressure ratio, efficiency, mass flow and energy addition. Stability is also a performance characteristic. It is linked to the response of the compressor to a disturbance that perturbs the compressor operation from a steady point. In transient disturbance, if the system returns to the original point of operational equilibrium, the performance is regarded as stable. The performance is considered unstable if the response is to drive operation away from the original point and steady state operation is not possible [7].

Moreover, there are two areas of compressor performance that relate to stability. One deals with operational stability and the other deals with aerodynamic stability. Operational stability is concerned with the matching of performance characteristics of the compressor with a downstream flow device such as a throttle, turbine or a jet nozzle.

It is common to see during the operations of the axial-flow compression systems, as the pressure rise increases, that the mass flow is reduced. A point will be reached when the pressure rise is a maximum. Further reduction in mass flow will lead to a sudden and definite change in the flow pattern in the compressor. Beyond this point, the compressor will enters into either a stall or a surge. Both stall and surge phenomena are undesirable and they can be detrimental in performance, structural integrity or system operations ([2], [3], and [7]).

In the area of instability caused by inlet distortion in axial compressor, there is a considerable interest over the years, with an extensive literature ([1], [3], [4], [5], [6], [8], [9] and [10]). Among them, Kim *et al.* [5] successfully calculated the qualitative trend of distorted performance and distortion attenuation of an axial compressor by using a simple integral method. Ng *et al.* [6] developed the integral method and proposed a distortion critical line. By making some simplifications, integral method can rapidly predict the distorted performance and distortion attenuation of an axial compressor without using comprehensive CFD codes and parallel supercomputer, and unavoidable, some elegance and detail of flow physics must be sacrificed. Nevertheless, the integral method can still provide a useful physical insight about the performance of the axial compressor with an inlet flow distortion.

In current work, using integral method, the behavior of the non-uniform inlet flow conditions in single and multistage axial compressor is studied. The distortion flow pattern through the compressor is also investigated. In addition, the off-line quality control method by Taguchi ([11] and [12]) is used to analyze the parameters affecting the flow through the compressor.

Kim *et al.* [5] has concluded that the two most important parameters to control the distortion propagation are the drag-to-lift ratio of the blade and the inlet flow angle. Taguchi method is used here to verify Kim *et al*'s findings on the parameters influencing the flow through the compressor.

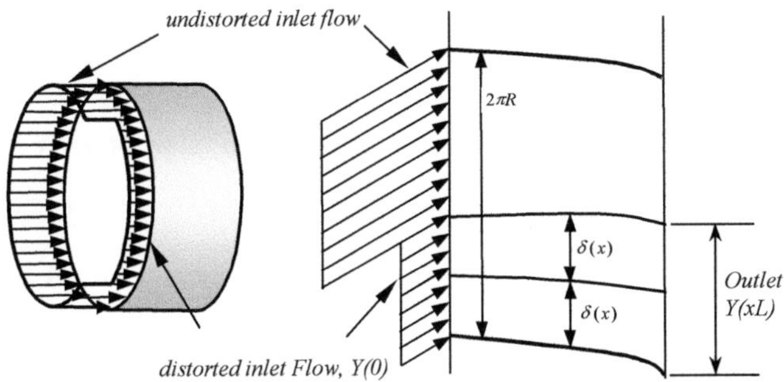

Fig. 3.1. A schematic of the distorted inlet flow and computational domain

3.2 Methodology

Consider a two-dimensional flow through a multistage compressor, as schematically shown in Fig. 3.1. In Kim *et al.*'s research, by integrating the 2-D inviscid Navier-Stokes equation, three ordinary differential equations, which describe the progress of flow in both the distorted and undistorted regions as it moves downstream in the machine, are derived.

$$\alpha \frac{d\alpha}{dx} = \frac{1}{K_3}\left(\frac{F_x - F_{x,0}}{U_0^{\,2}}\right) \tag{3.1}$$

$$\alpha \frac{d\beta}{dx} = \frac{1}{\overline{\gamma}}\left(\frac{F_y}{U_0^{\,2}}\right) \tag{3.2}$$

$$\alpha \frac{d\beta_0}{dx} = \frac{1}{\overline{\gamma}}\left(\frac{F_{y,0}}{U_0^{\,2}}\right) - \beta_0 K_2 \frac{d\alpha}{dx} \tag{3.3}$$

The axial and circumferential velocity components in undistorted region are $u = \alpha_0 U_0$; $v = \beta_0 V_0$, and in distorted region, $u = \alpha U_0$; $v = \beta V_0$, respectively. Here, U_0 and V_0 are axial and circumferential components of reference velocity at inlet, and $\overline{\gamma} = V_0 / U_0$. Where ($F_{x,0}$, $F_{y,0}$) and (F_x, F_y), denote the axial and circumferential forces in undistorted and distorted regions, respectively. The initial condition $\alpha_0(0)=\beta_0(0)=1$ is assumed. K_0 and K_1 are constants; K_2 and K_3 are the functions of $\alpha(x)$:

$$K_0 \equiv K_1 + (1 - K_1/\alpha)\alpha_0 \tag{3.4}$$

$$K_1 \equiv \frac{\delta\alpha}{\pi R} \tag{3.5}$$

$$K_2 = \frac{K_1(K_1 - K_0)}{(\alpha - K_1)^2} \tag{3.6}$$

$$K_3 = 1 + \frac{K_2(K_1 - K_0)}{\alpha - K_1} \qquad (3.7)$$

The distorted region size can be calculated following by the calculation of $\alpha(x)$ using (3.5):

$$Y(x) = \frac{\delta}{\pi R} = \frac{K_1}{\alpha} \qquad (3.8)$$

In general, their results show that the initial distorted region tends to grow as:

(i)   the drag-to-lift ratio increases;
(ii)   the upstream flow angle departs from the zero lift angle;
(iii)   the initial value of $\delta/\pi R$ increases; and
(iv)   the degree of the initial distortion of flow $[\alpha(0), \beta(0)]$ decreases.

And all these qualitative trends agree with intuitive anticipation.

   Taguchi method [11] is a very efficient tool for developing high quality products at a low cost. Using Taguchi methods for problem solving will:

(i)   provide a strategy for dealing with multiple and interrelated problems,
(ii)   give you a process that will provide a better understanding of your products and processes,
(iii)   give you a more efficient way of designing experiments for industrial problem solving using
(iv)   provide techniques for rational decision-making for prioritizing problems, allowing you to better focus your engineering resources, and
(v)   provide a tool for optimizing manufacturing processes.

Traditionally, one tends to change only one variable of an experiment at a time. The strength of the Taguchi technique is that one can change many variables at the same time and still retain control of the experiment. In present parametric research, we solve the integral equations, (3.1), (3.2), (3.3), by using orthogonal arrays, which is from Taguchi [12], to identify the factor/variable that has the most influence on the distortion so as to minimize it in the actual functioning of the axial compressor.

3.3 Results and Analysis

Investigations of distortion propagation conditions in axial compressors have been carried out. The main factors are identified: $\alpha(0)$ (inlet x-axis distorted velocity coefficient in the incompressible flow), θ_0 (inlet flow angle) and $K = k_D/k_L$ (the ratio of drag-to-lift coefficients). These parameters will be ranked according to their influence on the distortion using Taguchi table before subjected to further analysis. Three values are chosen for the inlet x-axis velocity coefficient in the distorted region: $\alpha(0)=$ 0.3, 0.5, and 0.7. The values chosen for the inlet flow angles are $\theta_0=15°$, 20°, and 25°, and the values of the ratio of drag-to-lift coefficients selected are $K=1.0$, 0.9, and 0.8, as shown in Table 3.1.

These values are chosen according to the research work done by Kim *et al.* [5]. So, by retaining these values in this project, the results obtained by Kim *et al.* can be verified and reaffirmed.

Table 3.1 Factors and levels

Factors	1	2	3
A	$\alpha(0)=0.3$	$\alpha(0)=0.5$	$\alpha(0)=0.7$
B	$\theta_0=15°$	$\theta_0=20°$	$\theta_0=25°$
C	$K=1.0$	$K=0.9$	$K=0.8$

Table 3.2 Layout on orthogonal array

	Factors			Parameters			Distorted Region
No.	A	B	C	$\alpha(0)$	θ_0	K	Size Y(xL) at xL=1.0
1	1	1	1	0.3	15	1.0	0.500407431923014
2	1	1	2	0.3	15	0.9	0.498290625981902
3	1	1	3	0.3	15	0.8	0.496132124051077
4	1	2	1	0.3	20	1.0	0.503192367192669
5	1	2	2	0.3	20	0.9	0.498846436096771
6	1	2	3	0.3	20	0.8	0.494418348585067
7	1	3	1	0.3	25	1.0	0.505553440992624
8	1	3	2	0.3	25	0.9	0.497321331316154
9	1	3	3	0.3	25	0.8	0.488924505958799
10	2	1	1	0.5	15	1.0	0.499480139318204
11	2	1	2	0.5	15	0.9	0.498393918048501
12	2	1	3	0.5	15	0.8	0.497296867242345
13	2	2	1	0.5	20	1.0	0.501553818573196
14	2	2	2	0.5	20	0.9	0.499432634631582
15	2	2	3	0.5	20	0.8	0.497288117161684
16	2	3	1	0.5	25	1.0	0.503425437636447
17	2	3	2	0.5	25	0.9	0.499696576056063
18	2	3	3	0.5	25	0.8	0.495938850940947
19	3	1	1	0.7	15	1.0	0.499254808571409
20	3	1	2	0.7	15	0.9	0.498789504826383
21	3	1	3	0.7	15	0.8	0.498323793294036
22	3	2	1	0.7	20	1.0	0.500458563750500
23	3	2	2	0.7	20	0.9	0.499527266782217
24	3	2	3	0.7	20	0.8	0.498595640370264
25	3	3	1	0.7	25	1.0	0.501586281747613
26	3	3	2	0.7	25	0.9	0.499979981856476
27	3	3	3	0.7	25	0.8	0.498378282945958

The three factors in Table 3.1 are arranged into an orthogonal array using three columns of $L_{27}(3^3)$, as shown on the left side of Table 3.2.

The experiments were carried out in 27 possible combinations as seen in Table 3.2. The numbers *1* to *27* on the left of the table are called experiment numbers.

For experiment No. 1 (row 1), as the number "1" appears in the orthogonal array for each of the factors A, B and C, it means that the experiment is calculated using factors in Table 3.1: $\alpha(0)=0.3$, $\theta_0=15°$ and $K=1.0$. Similarly, for experiment No. 6 (row 6), calculation is done using factors in Table 3.1: $\alpha(0)= 0.3$, $\theta_0=20°$ and $K=0.8$.

In Table 3.2, $Y(xL)$ is a representation of outlet distorted region size and it is non-dimensional. All cases examined have the same inlet distorted region size. $Y(0)$ is the inlet distortion region size and assumed to be *0.5.* where $xL=1$ means that it is a single stage compressor, while $x>10$ indicates it is a multistage compressor, and xL is the number of stages.

$Y(xL)$ is obtained by varying the values according to the analysis of variance (ANOVA). The *27* experimental cases provided a good comparison among the three parameters, $\alpha(0)$, θ_0 and K that are involved in the functioning of an axial compressor.

The distortion at the inlet, $Y(0)$, is *0.5* and it is used as a reference to measure the amount of $Y(x)$ at the outlet, $Y(xL)$. The amount of $Y(xL)$, i.e., $xL=1$, is tabulated in the rightmost column of Table 3.2.

The results produced from different values of $\alpha(0)$ are compared by the average value of $Y(xL)$ over the experiments that used $\alpha(0)=0.3$ (No. *1-9*), $\alpha(0)=0.5$ (No. *10-18*), and $\alpha(0)=0.7$ (No. *19-27*).

The average value of $Y(xL)$ is denoted by $\overline{Y}(xL)$, and they are:

$$\overline{Y}(xL)\,\big|_{\alpha(0)=0.3} = [0.500407431923014 + 0.498290625981902 + 0.496132124051077 +$$
$$0.503192367192669 + 0.498846436096771 + 0.494418348585067 +$$
$$0.505553440992624 + 0.497321331316154 + 0.488924505958799] \,/9$$
$$= 0.498120734677564$$

$$(3.9)$$

and:

$$\overline{Y}(xL)\,\big|_{\alpha(0)=0.5} = 0.499167373289885 \tag{3.10}$$

$$\overline{Y}(xL)\,\big|_{\alpha(0)=0.7} = 0.499432680460540 \tag{3.11}$$

Similarly, the results produced from different values of θ_0 are compared by the average value of $Y(xL)$ over the experiments using $\theta_0=15°$ (No. *1 - 3, 10 - 12, 19 - 21*), $\theta_0=20°$ (No. *4-6, 13-15, 22-24*), and $\theta_0=25°$ (No. *7-9, 16-18, 25-27*).

Likewise, the $\overline{Y}(xL)$ values for different K are compared in the same manner: $K=1.0$ (No. *1, 4, 7, 10, 13, 16, 19, 22, 25*), $K=0.9$ (No. *2, 5, 8, 11, 14, 17, 20, 23, 26*), and $K=0.8$ (No. *3, 6, 9, 12, 15, 18, 21, 24, 27*).

The above $\overline{Y}(xL)$ values are summarily put in Table 3.3 for comparison.

From the values of relative varying range in Table 3.3, among the different combinations of the values of various parameters ($\alpha(0)$, θ_0, K), the distortion is noticeably affected when the ratio of drag-to-lift coefficients of the blade (K) is varied. The percentage of distortion range is the least significant when θ_0 is varied, as compared to the results when other parameters are varied.

Hence, the parameter that has the most influence on the distortion is the ratio of drag-to-lift coefficients of the blade (K), followed by the x-axis distorted velocity coefficient in inlet ($\alpha(0)$). The inlet flow angle (θ_0) has the least influence on distortion as compared to the above two factors.

These results, however, contradict Kim *et al.*'s conclusion [5]. In their research, using only a few isolated cases, it was concluded that the two key parameters to control the growth of distortion propagation were the ratio of drag-to-lift coefficients of the blade and the angle of flow of the distorted upstream flow. There was no mention of the third parameter and ranking of these parameters was not carried out.

However, upon using Taguchi off-line quality control method in this research, the parameters are ranked according to their degree of influence in distortion.

From the results obtained from Taguchi method, when the inlet x-axis velocity coefficient, $\alpha(0)$, is *0.7*, it has the lowest value of the increment of distortion region size, the difference between $Y(xL)$ and $Y(0)$, $\Delta Y = Y(xL)-Y(0)$, at the outlet as compared to the other two values: *0.3* and *0.5*. Therefore, $\alpha(0)$ is chosen to be *0.7* in the calculations for further analysis of graphs.

Three areas of studies were carried out and categorized into three case studies:

Case study 1: drag-to-lift ratio, K, is varied,
Case study 2: x-axis inlet distorted velocity coefficient, $\alpha(0)$, is varied, and
Case study 3: inlet flow angle, θ_0, is varied.

Table 3.3 Summarized results of the various factors

Parameter	Value	Average Y(xL) at xL=1.0	Range (%)
$\alpha(0)$	0.3	0.498120734677564	
	0.5	0.499167373289885	0.131194578297539
	0.7	0.499432680460540	
θ_0	15°	0.498485468139652	
	20°	0.499257021460439	0.077155332078666
	25°	0.498978298827898	
K	1.0	0.501656921078409	
	0.9	0.498919808399561	0.551286212838892
	0.8	0.496144058950020	

In following cases and analysis, for ease in mentioning, we use $Y(x)$ with $x=1$ and $x=10$ to express the outlet distorted region size for single-stage and ten-stage compressors, respectively.

3.3.1 Case Study 1: Drag-to-lift Ratio, *K*, is Varied

Shown in Fig. 3.2 is a comparison of two graphs for $Y(x)$ ($x=1$ and $x=10$), with $\alpha(0)$, θ_0 kept constant at *0.7* and *15°* respectively. The only parameter varied is the drag-to-lift ratio, from *0.5* to *1.5*. A multistage compressor of *10* stages is chosen for comparison because the trend of outlet distortion region size for x more or less than ten is similar to $x=10$.

For the graph of $x=10$, it is observed that as K increases, the distortion region grows upstream, and the difference in the inlet and outlet distortion region size $(\Delta Y=Y(x)-0.5)$ reduces. As the gradient of propagation of multistage compressor grows steeper than that of single stage compressor, the value of ΔY for a single stage compressor is noted to be much smaller than that of a multistage.

In other words, a multistage compressor has a larger outlet distortion region size than a single-stage compressor when the drag-to-lift ratio increases.

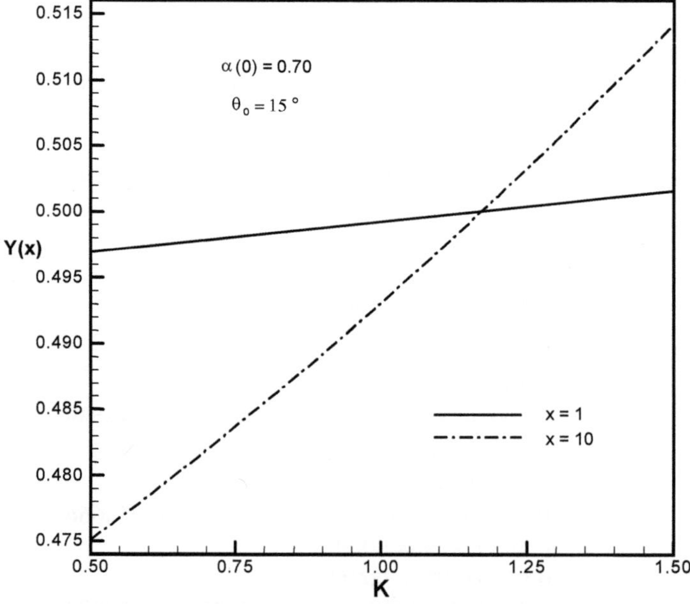

Fig. 3.2. The effects of drag-to-lift ratio on the outlet distortion region size for single stage and multistage compressors, $\alpha(0)=0.7$, $\theta_0=15°$

Figure 3.3 shows the trend of propagation for $x = 1$ and $x = 10$, where the inlet flow angle is 20° in this case. A multistage compressor of 10 stages is again chosen for comparison.

Similar observations are seen, as in Fig. 3.2, where the difference between the inlet and outlet distortion region sizes ($\Delta Y= Y(x)-0.5$) reduces with increasing value of K. As the gradient of propagation of $x=10$ grows steeper than that of $x=1$, the value of ΔY for a single stage compressor is also noted to be much smaller than that of a multistage.

Here, larger flow angle is noted to cause a growing effect on the propagation of distortion, where the difference of the inlet and outlet distortion region size is positive, i.e., $Y(x) - Y(0) > 0$.

Figure 3.4 depicts the propagation of distortion for a single stage compressor. In this case, x is taken to be 1 and the flow angles are $15°$ and $20°$. There is no variation at all in the propagation trends between $\theta_0 = 15°$ and $20°$. They superimposed on each other. Hence, it can be concluded that for a single-stage compressor, the inlet flow angle has no significant effect on the propagation of distortion at all.

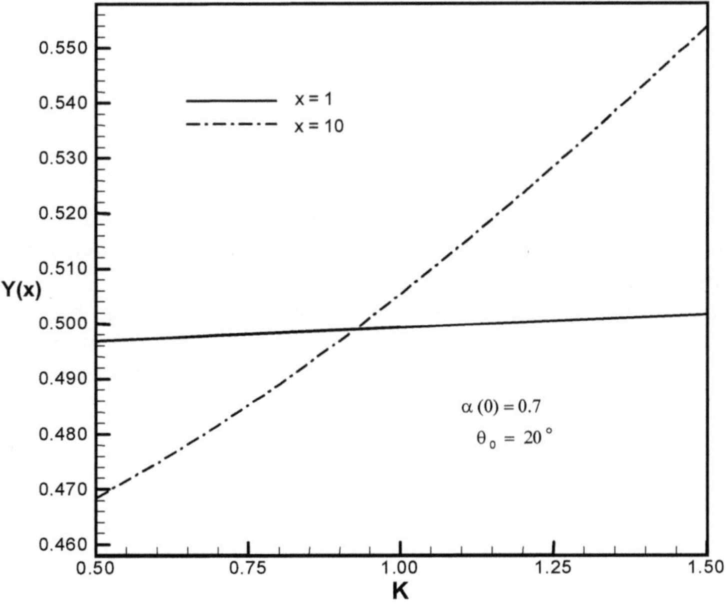

Fig. 3.3. The effects of drag-to-lift ratio on the outlet distortion region size for a single stage and multistage compressor, $\alpha(0)=0.7$, $\theta_0=20°$

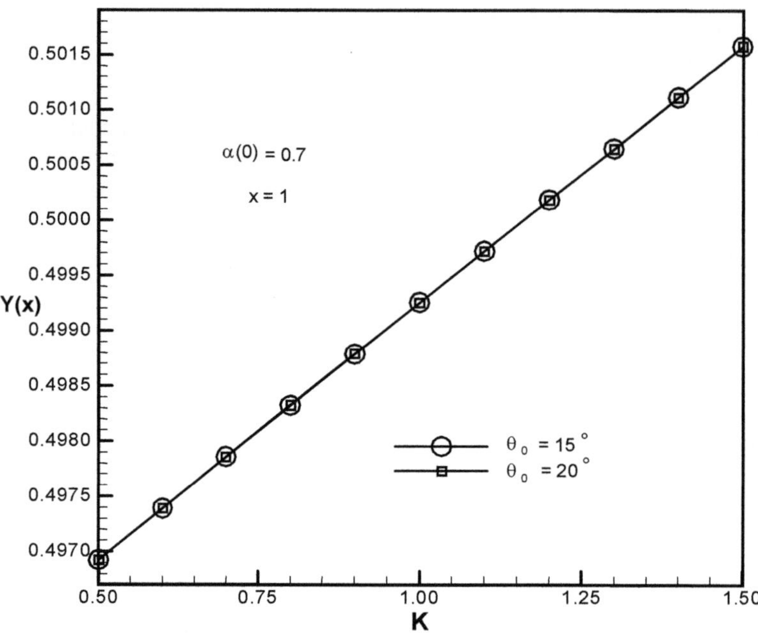

Fig. 3.4. The effects of drag-to-lift ratio on the outlet distortion region size for a single stage compressor, $\alpha(0)=0.7$, $\theta_0=15°$ and $\theta_0=20°$

The propagation of distortion for a multistage compressor is shown in Fig. 3.5, with $x=10$. Although the distortion region grows for both plots, a significant difference in their sizes is noted for $\theta_0 = 15°$ and $20°$.

Keeping the x-axis inlet distorted velocity coefficient constant, it can be seen that a larger ratio of drag-to-lift coefficient with a larger angle of flow of the upstream flow could cause a higher increase of distortion propagation for a multistage compressor.

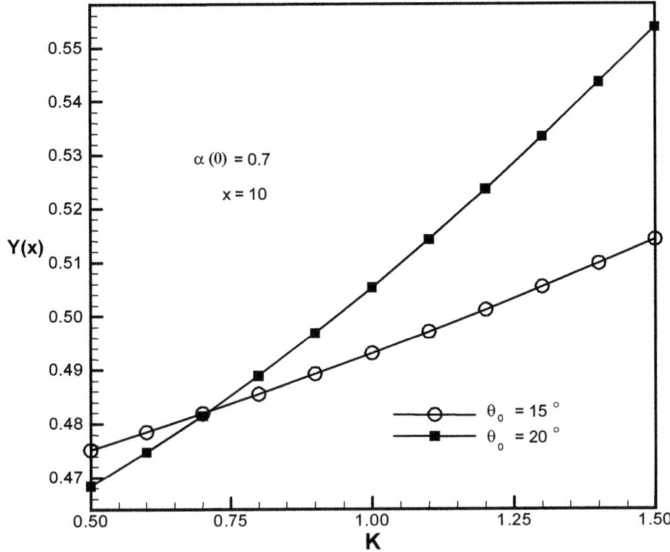

Fig. 3.5. The effects of drag-to-lift ratio on the outlet distortion region size for a multistage compressor, $\alpha=0.7$, $\theta_0=15°$ and $\theta_0=20°$

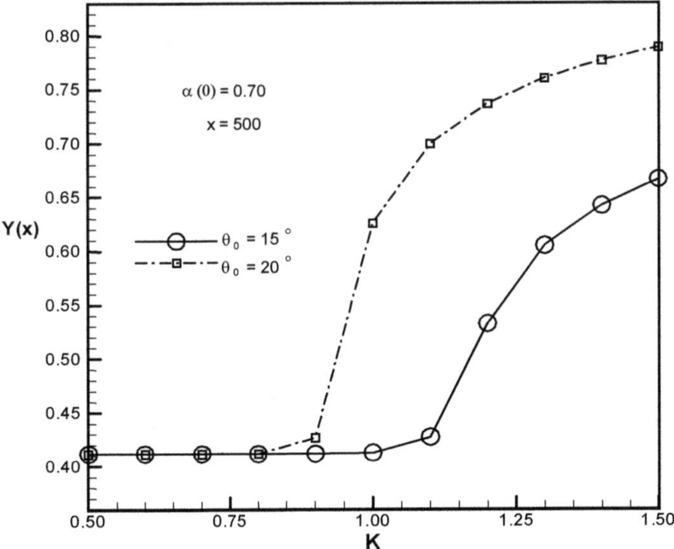

Fig. 3.6. The asymptote behavior of drag-to-lift ratio on the outlet distortion region size for a high stage compressor, x=500. $\alpha(0)=0.7$

In order to show the asymptotic behavior of the inlet distortion propagation, a virtual compressor with 500 stages is analyzed. Figure 3.6 illustrates its trends of the distortion propagation when $\theta_0=15°$ and $\theta_0 =20°$. It is noted that there is no significant difference between the two plots when the ratio of drag-to-lift, K, is small. However, with increasing K, a larger inlet flow angle could result in higher distortion propagation for a multistage compressor. Of course, practically, it is presently impossible for a compressor to have 500 stages in it, hence, this figure is only meant for illustrative purpose of showing the propagation pattern.

3.3.2 Case Study 2: *X*-axis Inlet Distorted Velocity Coefficient, α(0), is Varied

In Fig. 3.7 and Fig. 3.8, the inlet distortion propagation trends for all values of *α(0)* are similar for a single-stage compressor (*x=1*), and a multistage compressor (*x=10*). Both figures show how the distortion propagates as the flow progresses downstream for the case of inlet flow angle $\theta_0 =15°$.

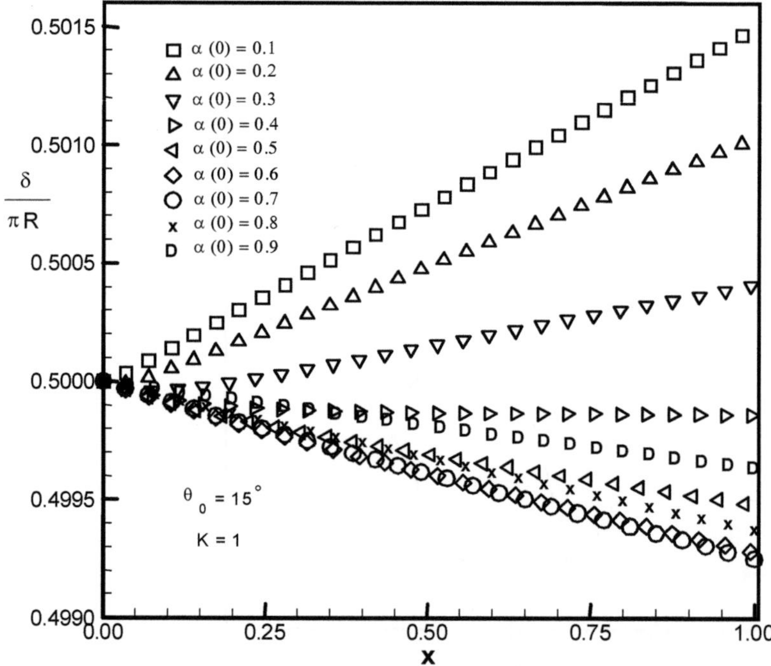

Fig. 3.7. The effect of degree of the initial distortion for a single stage compressor, $\theta_0=15°$, $K=1.0$

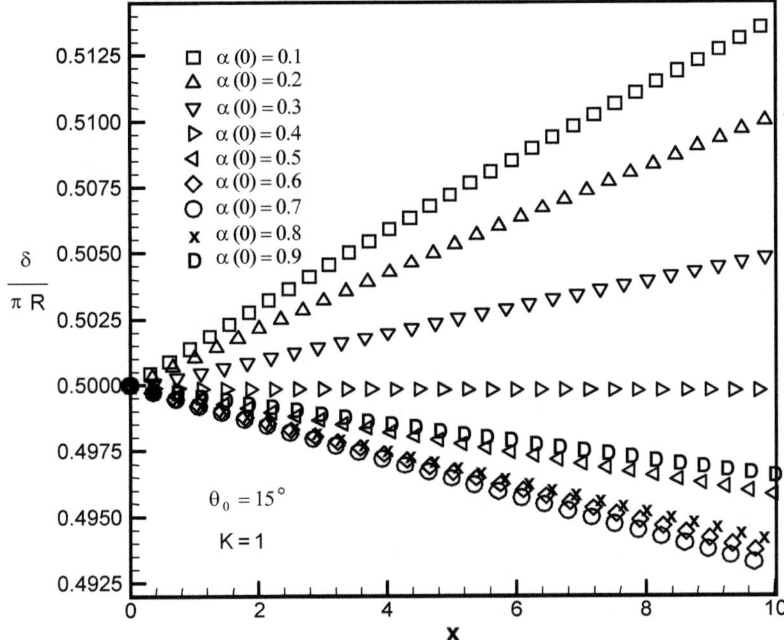

Fig. 3.8. The effect of degree of the initial distortion for a multistage compressor, $\theta_0=15°$, $K=1.0$

It can be seen that for smaller values of $\alpha(0)$, the outlet distorted region size increases. For $\alpha(0)>0.7$, the increment of distorted region size between the inlet and the outlet region ($\Delta Y=Y(x)-0.5$) actually decreases as compared with the values when $\alpha(0)$ is smaller or equal to 0.7. Remembering the assumption made, that $\alpha_0(0)=1$ and the initial distortion region size is 0.5, these curves actually tend toward the $Y(x)$-value: 0.5—no distortion when $\alpha(0)\to1$.

Hence, it can be concluded that the distorted region with a large initial distortion will grow while that with a smaller distortion will tend to lose its ground.

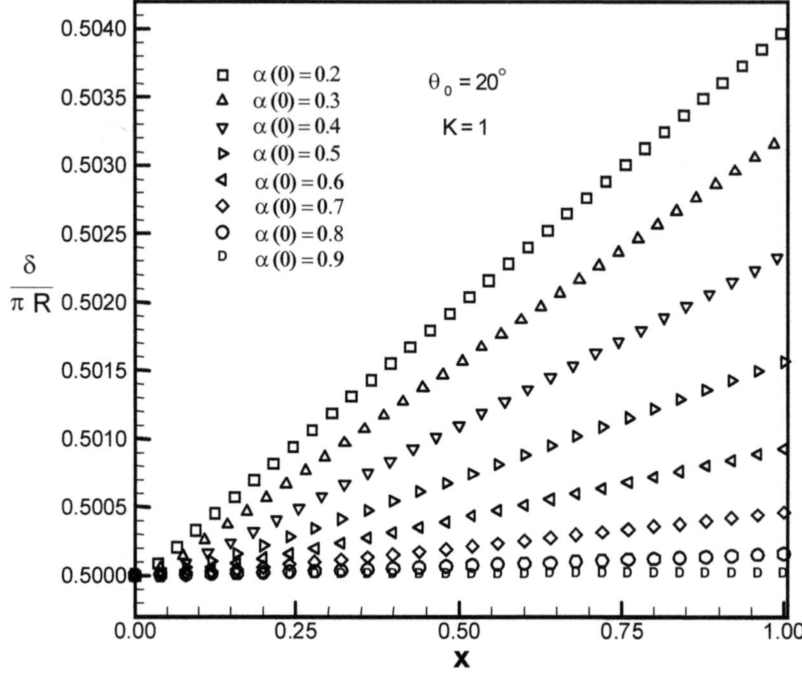

Fig. 3.9. The effect of degree of the initial distortion for a single stage compressor, $\theta_0=20°$, $K=1.0$

Next, the influence of the $\alpha(0)$ on the distortion propagation with $\theta=20°$ is shown in Fig. 3.9 and Fig. 3.10. With a difference in the flow angle used, similar distortion propagation are observed in the two figures for a single-stage compressor, $x=1$, and a multistage compressor, $x = 10$. It is seen that for smaller values of $\alpha(0)$, the outlet distorted region size increases.

It can also be noted that at a high inlet flow angle ($\theta_0=20°$), all inlet distorted velocity coefficients will tend to make the distorted region grow downstream, despite whether it is a single stage compressor or a multistage compressor. Only when the inlet distorted velocity coefficient tends to unity (no distortion), the distorted region tends to no variety ($Y(x)\equiv0.5$).

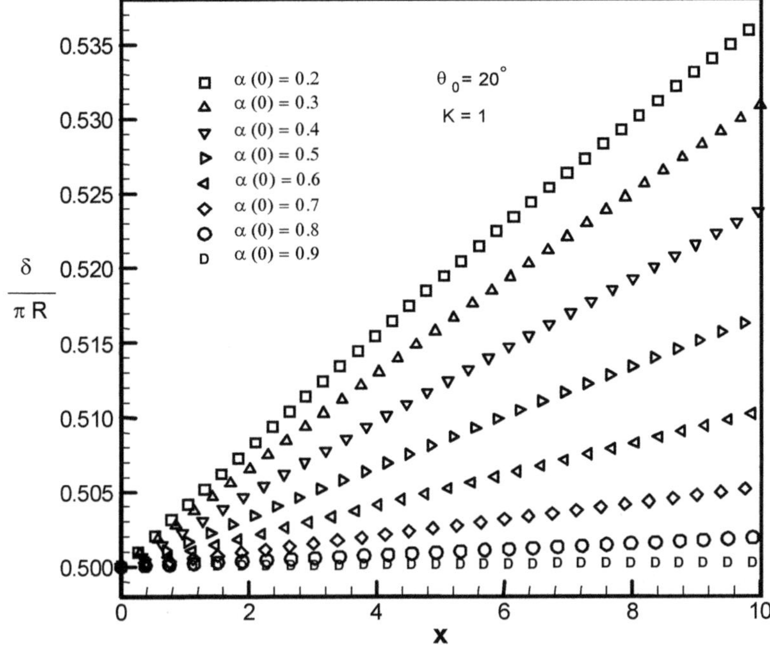

Fig. 3.10. The effect of degree of the initial distortion for a multistage compressor, $\theta_0=20°$, $K=1.0$

Again, Fig. 3.11 is used for illustrative purpose for a large number of stages ($x=500$). This figure uses values of $\alpha(0)$ from *0.3* to *0.9*.

As seen in Fig. 3.11, for the conditions investigated, the distorted regions first grow but eventually level off far downstream. It can be noted that smaller values of $\alpha(0)$ would result in larger amount of outlet distorted region size. Keeping the flow angle at *20°* and the drag-to-lift ratio at *1.0*, a smaller $\alpha(0)$ results in an asymptotic behavior on the distortion propagation. The distorted region cannot occupy the entire region as it will violate the continuity but tend to reach equilibrium itself with the undistorted region through the mass continuity.

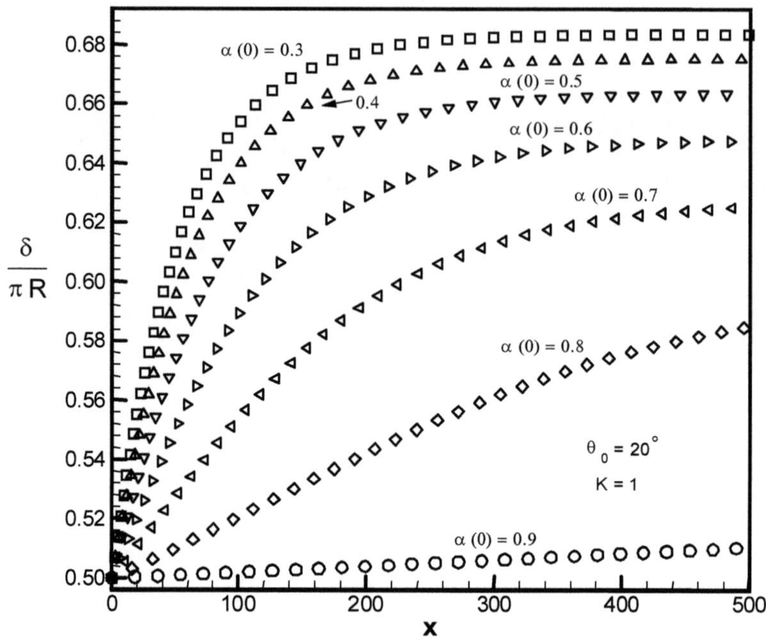

Fig. 3.11. The asymptotic behavior of the inlet distortion for a high-stages compressor, $x=500$, $\theta_0=20°$, $K=1.0$

3.3.3 Case Study 3: Inlet Flow Angle, θ_0, is Varied

In Fig. 3.12 and Fig. 3.13, the distortion propagation are similar for a single stage compressor and a multistage compressor. By keeping the x-axis inlet distorted velocity coefficient, $\alpha(0)=0.7$, and the drag-to-lift ratio, $K=1.0$, constant, the distorted inlet flow region is seen to expand as it moves downstream for high angles of flow, $\theta_0=19°$ to $22°$, but diminishes for low angle of flow such as $\theta_0=12°$ to $17°$. The outlet distorted region size is larger with higher angle of flow in either a single-stage compressor or a multistage compressor.

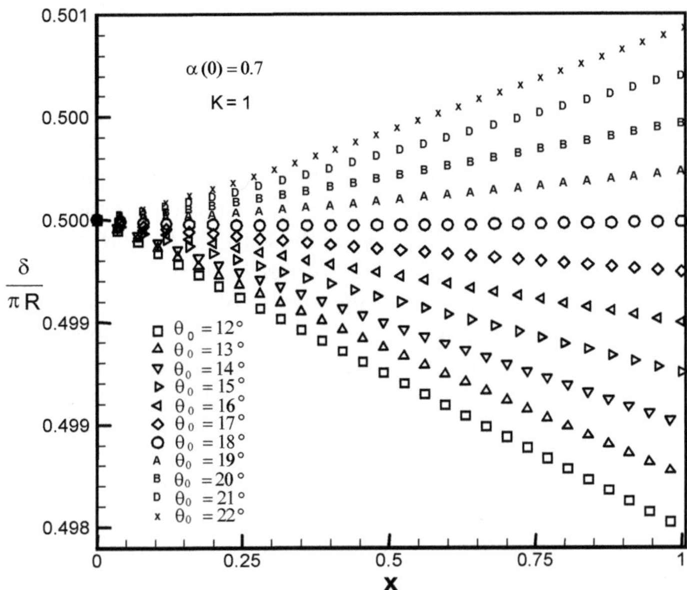

Fig. 3.12. The effect of the angle of flow for a single stage compressor, $\alpha(0)=0.7$, $K=1.0$

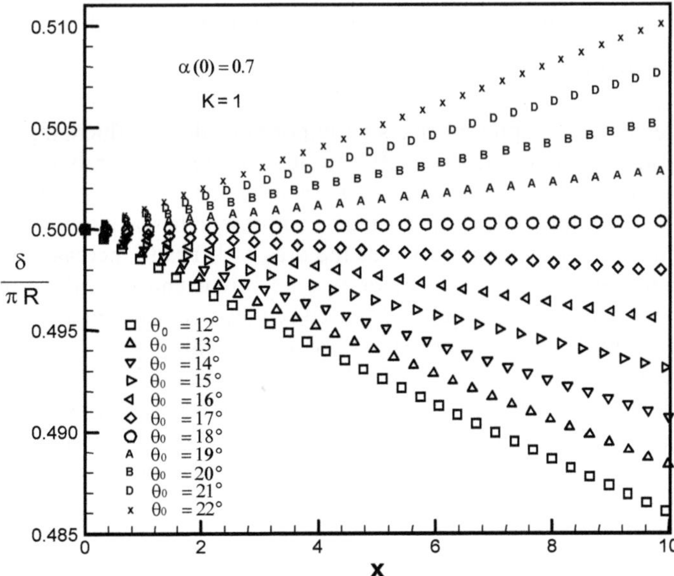

Fig. 3.13. The effect of the angle of flow for a multistage compressor, $\alpha(0)=0.7$, $K=1.0$

3.4 Conclusion

In this work, the parameters investigated are the ratio of drag-to-lift coefficients of the blade, K, the x-axis inlet distorted velocity coefficient, $\alpha(0)$, and the angle of flow of the undistorted upstream flow, θ_0.

Upon application of Taguchi off-line quality control method, these parameters are ranked according to their degree of influence on the distortion propagation in an axial compressor. The ranking, from the most to the least, is: K, $\alpha(0)$, and θ_0.

These results differ from the conclusion that Kim *et al.* had made. Stated in their article, the two key parameters to control the growth of the distortion propagation are the ratio of drag-to-lift coefficients and the angle of flow of the undistorted upstream flow. However, the method they used was only integral method and found out their conclusion from several cases, which is questionable. Apparently, Taguchi method used in current work is the right approach to provide an exact ranking result for these affecting parameters.

The current work investigates not only on the integral method, but also take into the account of the conditions of the product when it encounters disturbance when in use. This may in turn affect the company's reputation in the market. Hence, off-line quality control engineering is used here to verify the parameters.

In general, the results from this research show that the initial distorted region tend to grow when:

(i)   drag-to-lift ratio increases;

(ii)   inlet distorted velocity coefficient of the compressor decreases; and

(iii)   angle of flow increases.

The number of stages in the compressor plays an important role in influencing the distortion propagation pattern. The drag-to-lift ratio has more significantly effects on the inlet distortion propagation when the comparison is done between a single-stage and a multistage compressor. A multistage compressor has a larger outlet distortion size than a single-stage compressor as the drag-to lift ratio increases. On the other hand, when the inlet distorted velocity coefficient and the inlet flow angle are altered, the difference in the outlet distorted region size between a single-stage compressor and a multistage compressor is less significant.

It would be more appropriate to improve on the airfoil characteristics to obtain more accurate calculations. Current integral method assumed blade lift vanishes when $\theta=\theta^*$ and drag is zero at this flow angle. Here θ and θ^* are local angle of flow and reference angle, respectively; and C_l and C_d are the general lift and drag coefficients used in wing theory.

$$C_l = k_L\,(tg\theta - tg\theta^*)\tag{3.12}$$

$$C_d = k_D\,(tg\,\theta - tg\,\theta^*)\tag{3.13}$$

such that

$$F_\perp = k_L \frac{1}{2}(u^2 + v^2)(tg\theta - tg\theta^*) \qquad (3.14)$$

$$F_\parallel = k_D \frac{1}{2}(u^2 + v^2)(tg\theta - tg\theta^*)^2 \qquad (3.15)$$

It would be encouraged to use experimental wing section data and curve fitting on wings, for example NACA 65 series [13], to obtain an improved version of the coefficients in functions of angle of attack:

$$C_l = f_1(angle\ of\ attack) \qquad (3.16)$$

$$C_d = f_2(C_l) \qquad (3.17)$$

These data can build up expressions of the coefficients of lift and drag of the blade and in turn, provide more suitable force components for more accurate calculations.

References

[1]   Chue, R., Hynes, T.P., Greitzer, E.M., Tan, C.S., and Longley, J.P., 1989, Calculations Of Inlet Distortion Induced Compressor Flow Field Instability, *International Journal of Heat and Fluid Flow*, **10(3)**: 211-223.

[2]   Cumpsty, N.A., 1989, Compressor Aerodynamics, Harlow, Essex, England: Longman Scientific and Technical; New York: J. Wiley.

[3]   Greitzer, E.M., 1980, Review-Axial Compressor Stall Phenomena, *ASME Journal of Fluids Engineering*, **102**: 134-151.

[4]   Greitzer, E.M. and Griswold, H.R., 1976, Compressor-Diffuser Interaction With Circumferential Flow Distortion, *Journal of Mechanical Engineering Science*, **18(1)**: 25-38.

[5]   Kim, J.H., Marble, F.E., and Kim, C.–J., 1996, Distorted Inlet Flow Propagation In Axial Compressors, In *Proceedings of the 6th International Symposium on Transport Phenomena and Dynamics of Rotating Machinery*, **2**: 123-130.

[6]   Ng, E.Y-K., Liu, N., Lim, H.N. and Tan, T.L., 2002, Study On The Distorted Inlet Flow Propagation In Axial Compressor Using Integral Method, *J. Computational Mechanics*, **30(1)**: 1-11.

[7]   Pampreen, R.C., 1993, "Compressor Surge and Stall", USA: Concepts ETI. Inc.

[8]   Reid, C., 1969, The Response Of Axial Flow Compressors To Intake Flow Distortion, In *Proceeding of International Gas Turbine and Aeroengine Congress and Exhibition*, ASME Paper 69-GT-29.

[9]   Stenning, A.H., 1980, Rotating Stall And Surge, *Journal of Fluids Eng.*, **102**: 14-20.

[10]   Stenning, A.H., 1980, Inlet Distortion Effects In Axial Compressors, *Journal of Fluids Eng.*, **102**: 7-13.

[11] Taguchi, G., 1993, Taguchi On Robust Technology Development: Bringing Quality Engineering Upstream, New York: ASME Press.

[12] Taguchi, G., 1986, Introduction To Quality Engineering: Designing Quality Into Products And Processes, Japan: Asian Productivity Organization.

[13] Von Doenhoff, E.A., 1959, Theory Of Wing Sections, Including A Summary Of Airfoil Data, New York, Dover Publications.

Chapter 4
A Development of Novel Integral Method for Prediction of Distorted Inlet Flow Propagation in Axial Compressor

Based on the original integral method and its applications mentioned and discussed in previous chapters, an improved integral method is proposed and developed for the quantitative prediction of distorted inlet flow propagation through axial compressor. The novel integral method is formulated using more appropriate and practical airfoil characteristics, with less assumption needed for derivation. The results indicate that the original integral method [6] underestimated the propagation of inlet flow distortion. The effects of inlet flow parameters on the propagation of inlet distortions, as well as on the compressor performance and characteristic are simulated and analyzed. From the viewpoint of compressor efficiency, the propagation of inlet flow distortion is further described using a compressor critical performance and its associated critical characteristic. The results present a realistic physical insight to an axial flow compressor behavior with a propagation of inlet distortion.

4.1 Introduction

Compression system is an important component of the gas turbine engine, and its performance strongly influences the performance of all other components. The operational envelopes of modern compressors also demand a challenging trade-off between safety and performance due to the inherent aerodynamic instabilities associated with compressor stall. Various dynamic events such as distorted inlet flow in axial compressor, rotor blade tip clearance changes, rapid changes in the operating conditions (fuel throttling) may cause the component-mismatch and instability problems including rotating stall or even surge (oscillations in the mass flow rate), or a combination of them, which might result in catastrophic damage to the entire engine. Engines are thus constrained to operate below the surge line termed as a safe surge margin. Because the safe surge margin must accommodate the most extreme events, the surge margin is therefore sizable and causes the compressor to operate below the optimal conditions. The ability to detect and prevent an impending stall allows the engine to be designed and safely operated at maximum efficiency without sacrificing engine weight in future gas turbine engine development. The operability of current and future gas turbine engines can then be greatly enhanced.

To understand and avoid compressor instability due to flow distortion, the inlet distortion and its propagation effects have received much attention over the years. The early work about analytical and experimental descriptions of propagating stall can be traced back to mid-fifties. In this period, one of remarkable work was done by Marble [8]. Marble proposed a simple model to yield essential features of stall propagation, such as dependence on the extent of stalled region upon operating conditions, the pressure loss associated with stall, and the angular velocity of stall propagation. In the work done by Emmons et al. [3], an experimental investigation was performed to verify their theory of instability about the phenomena in surge and stall propagation. More work had been done in the recent years. Cumpsty and Greitzer [2] proposed a simple model for compressor stall cell propagation. Jonnavithula et al. [5] presented a numerical and experimental study of stall propagation in axial compressors. Longley et al. [7] described stability of flow through multistage axial compressors, and so on.

As to the work focused on the analyses of distorted inlet flow, Reid [13] presented an insight into the mechanism of the compressor's response and tolerance to distortion. One of the methods used in the analyses was a linearized approach ([12] and [4]), that provides a quantitative information about the performance of the compressor in a circumferentially non-uniform flow. Several models ([10] and [4]), such as the parallel compressor model and its extensions [9], were used to assess the compressor stability with inlet distortion. Stenning [15] also presented some simpler techniques for analyzing the effects of circumferential inlet distortion.

The numerical simulation of complex flows within multiple stages of turbomachinery is becoming more effective and is useful for design application today by using the advanced computers. However, a large-scale simulation with CFD codes still requires huge computing resources far exceeding the practical limits of most single-processor supercomputers. Many CFD codes have to be performed on a parallel supercomputer ([1] and [17]). In order to rapidly predict the distorted performance and distortion attenuation of an axial compressor without using comprehensive CFD codes and parallel supercomputer, it is necessary to make some simplifications, and some elegance and detail of flow physics must be sacrificed. Kim et al. [6] successfully calculated the qualitative trend of distorted performance and distortion attenuation of an axial compressor by using an overly simplified integral method. Instead of solving a detailed flow field problem, the integral method renders the multistage analysis as a nature part, and permits large velocity variations, including back flow. Ng et al. [11] developed the integral method and proposed a distortion critical line. The integral method provided a useful physical insight about the performance of the axial compressor with an inlet flow distortion. It is thus meaningful to further develop and refine this method.

In the present study, the authors further improve and develop the integral method from the previous one [6] by adopting more appropriate and realistic airfoil characteristics. The calculated results indicate that the previous integral method underestimated greatly the inlet distortion propagation. By using the newly developed integral method, an investigation is proceeded to present the

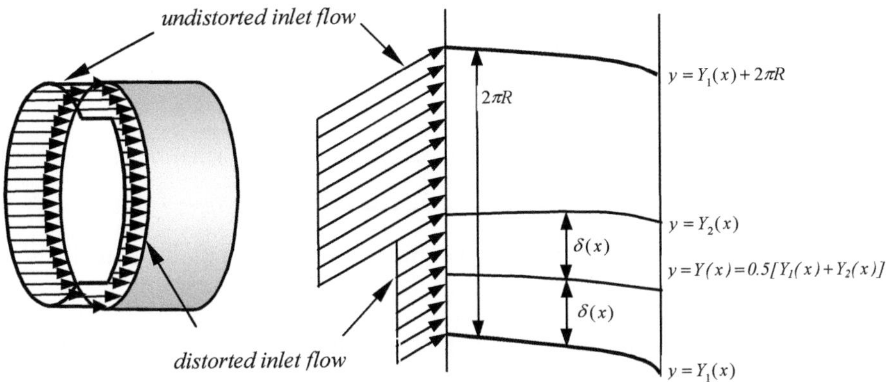

Fig. 4.1. The distorted inlet flow and its two-dimensional schematic used for integral method

effects of inlet parameters upon the downstream flow features with inlet distortion, including the inlet distortion propagation, the compressor critical performance and critical characteristic. The airfoil characteristics are also derived and discussed.

4.2 Theoretical Formulation

Consider a two-dimensional inviscid flow through an axial compressor as shown in Fig. 4.1. In the x, y plane, the circumferential extent of compressor is denoted by y –direction and its period is $2\pi R$. The flow field is divided as two parts, one is undistorted flow and another is the distorted flow. The distorted flow moves along a line $y = Y(x)$ and extends a distance $\delta(x)$ on each side of this line to form a stream tube. Thus, the distorted region in y-direction is from $y = Y(x) - \delta(x)$ to $y = Y(x) + \delta(x)$, and the undistorted region is from $y = Y(x) + \delta(x)$ to $y = 2\pi R + Y(x) - \delta(x)$.

The flow that is subjected to a force field can be described by the equations of continuity and motion:

$$\frac{\partial u}{\partial x} + \frac{\partial v}{\partial y} = 0 \tag{4.1a}$$

$$\frac{\partial(u^2)}{\partial x} + \frac{\partial(uv)}{\partial y} + \frac{\partial}{\partial x}(\frac{p}{\rho}) = F_x \tag{4.1b}$$

$$\frac{\partial(uv)}{\partial x} + \frac{\partial(v^2)}{\partial y} + \frac{\partial}{\partial y}(\frac{p}{\rho}) = F_y \tag{4.1c}$$

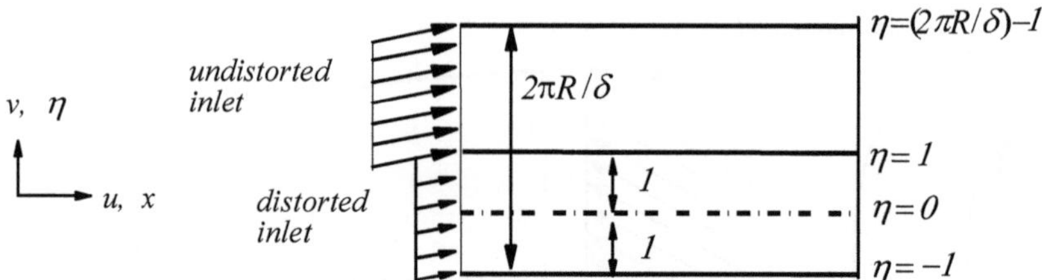

Fig. 4.2. A schematic of coordinates transformation on computational domain

For simplifying the derivation of integral equations, the coordinates system is transformed in the circumferential direction from (x, y) to (x, η) using $\eta = \dfrac{y - Y(x)}{\delta(x)}$. The computational domain is thus transformed into a parallel channel, as shown in Fig. 4.2.

The integral technique is to integrate (4.1), with respect to y, and to determine the development of the flow from the inlet toward the downstream.

Before integrating the equations of motion and continuity in the distorted region, undistorted region and overall region respectively, the matching of velocity profiles and pressure field should be chosen.

To illustrate the procedure of integral method, let consider a simple example.

The velocities in each of the regions are taken to be independent of y whereas the velocity components in each of the regions are defined as:

$$u = \alpha(x)U_0 \qquad (4.2a)$$

$$v = \beta(x)V_0 \qquad (4.2b)$$

$$u_0 = \alpha_0(x)U_0 \qquad (4.2c)$$

$$v_0 = \beta_0(x)V_0 \qquad (4.2d)$$

The inlet velocity has an flow angle of θ_0, and:

$$\overline{\gamma} = tan\,\theta_0 = V_0/U_0 \qquad (4.3)$$

where U_0 and V_0 are the x- and y- components of reference inlet velocity respectively. The distorted velocity coefficients $\alpha(x)$ and $\beta(x)$ are the velocity fractions of the referenced inlet velocity in the distorted inlet region, and the undistorted velocity coefficients $\alpha_0(x)$ and $\beta_0(x)$ are the velocity fractions of the referenced inlet velocity in the undistorted inlet region respectively. u and v are the x- and y- components of distorted velocity, while u_0 and v_0 are the x- and y- components of undistorted velocity respectively.

The static pressure is taken circumferentially (vertically) uniform:

$$\frac{p}{\rho} \equiv \frac{p}{\rho}(x) \tag{4.4}$$

This assumption is taken in simplifying the process of derivation, which means that fluid pressure in the distorted region is fixed along y-direction, and this assumption would cause an overestimation of distortion level and propagation. In other words, if the decrease of static pressure is neglected, the inlet-distorted velocity that is calculated from the measured-inlet total pressure will be smaller than the reality. Therefore, the difference of the velocities between two regions at inlet will be increased compare with the reality, and the inlet distortion would be overestimated. The predicted results of propagation of distortion would be larger than the real one, and hence provide a wider safety margin. The more the compressor's stage number (longer axial scale) is, the higher would be the magnitude of overestimated distortion.

In the equations of motion and continuity, the force components F_x and F_y are the replacements of the equivalent terms acting on the blades as shown in Fig. 4.3. The multi-stages therefore become a natural part of compressor. The following are the definitions of the forces acting on the stator and rotor in the region of

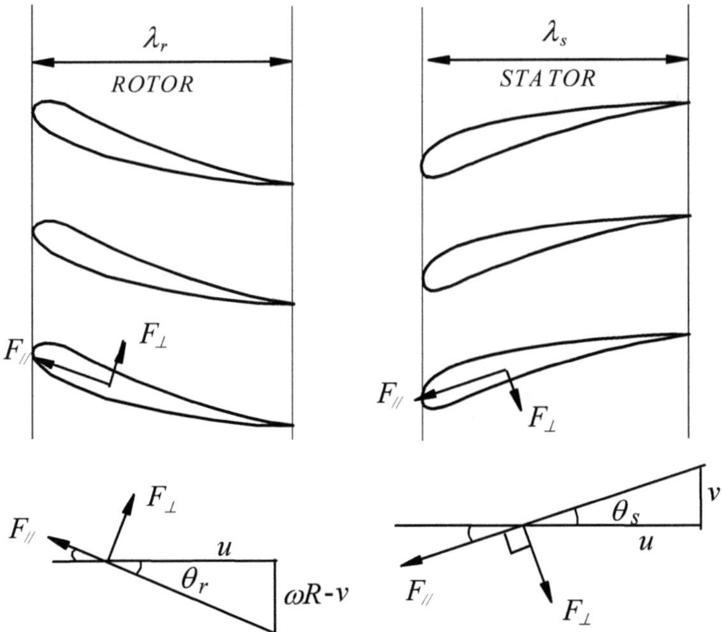

Fig. 4.3. The force diagram and velocities in the compressor stage

distortion according to the blade element theory (replacing u and v by u_0 and v_0 in the undistorted region):

$$F_\perp^{\ s} = \frac{C_l}{2}(u^2 + v^2)$$
(4.5a)

$$F_{//}^{\ s} = \frac{C_d}{2}(u^2 + v^2)$$
(4.5b)

$$F_\perp^{\ r} = \frac{C_l}{2}[u^2 + (\omega R - v)^2]$$
(4.5c)

$$F_{//}^{\ r} = \frac{C_d}{2}[u^2 + (\omega R - v)^2]$$
(4.5d)

Here, C_l and C_d are the lift and drag coefficients respectively. The force components for a unit circumferential distance of a complete stage are:

$$F_x = \frac{\lambda_r}{\overline{\lambda}}\left(F_\perp^{\ r} \sin\theta_r - F_{//}^{\ r} \cos\theta_r\right) + \frac{\lambda_s}{\overline{\lambda}}\left(F_\perp^{\ s} \sin\theta_s - F_{//}^{\ s} \cos\theta_s\right)$$
(4.6a)

$$F_y = \frac{\lambda_r}{\overline{\lambda}}\left(F_\perp^{\ r} \cos\theta_r + F_{//}^{\ r} \sin\theta_r\right) + \frac{\lambda_s}{\overline{\lambda}}\left(- F_\perp^{\ s} \cos\theta_s - F_{//}^{\ s} \sin\theta_s\right)$$
(4.6b)

where the superscript and subscript s denote the stator; the superscript and subscript r denote the rotor; the $\lambda_r/\overline{\lambda}$ and $\lambda_s/\overline{\lambda}$ are the relative length of rotor and stator in a single stage respectively. Here, we assume that $\overline{\lambda} = \lambda_r + \lambda_s$. The angles θ_s and θ_r are the local flow angles with respect to stator and rotor. For example, in distorted region,

$$\tan\theta_s = v/u$$
(4.7a)

$$\tan\theta_r = (\omega R - v)/u$$
(4.7b)

The (u, v) will be replaced by (u_0, v_0) in undistorted region.

With substitution of (4.2), (4.3), (4.5), (4.7) into (4.6), and simplifying the resulted equations by using $\sigma = \dfrac{\omega R}{V_0}$, we obtain:

$$\frac{F_x}{U_0^2} = \frac{\lambda_r}{2\overline{\lambda}\sigma^2\overline{\gamma}^2}[C_l(\sigma - \beta)\overline{\gamma} - C_d\alpha]\sqrt{\alpha^2 + (\sigma - \beta)^2\overline{\gamma}^2} + \frac{\lambda_s}{2\overline{\lambda}\sigma^2\overline{\gamma}^2}(C_l\beta\overline{\gamma} - C_d\alpha)\sqrt{\alpha^2 + \beta^2\overline{\gamma}^2}$$
(4.8a)

$$\frac{F_y}{U_0^2} = \frac{\lambda_r}{2\overline{\lambda}\sigma^2\overline{\gamma}^2}[C_l\alpha + C_d(\sigma - \beta)\overline{\gamma}]\sqrt{\alpha^2 + (\sigma - \beta)^2\overline{\gamma}^2} - \frac{\lambda_s}{2\overline{\lambda}\sigma^2\overline{\gamma}^2}(C_l\alpha + C_d\beta\overline{\gamma})\sqrt{\alpha^2 + \beta^2\overline{\gamma}^2}$$
(4.8b)

Similarly, the force components in the undistorted region $F_{x,0}$ and $F_{y,0}$ can be obtained using expression of undistorted velocity coefficients α_0 and β_0. Here, the equations for the force components are different from the previous expressions [6]. The current effort is derived by adopting an exact blade element theory. On the other hand, unlike the oversimplified procedure with the constants lift and drag coefficients, the more practical coefficients are to be applied according to the experimental results of airfoil sections [16].

In the distorted region, because the boundaries are streamlines, we obtain the following equations by integrating (4.1) along η-direction:

$$\delta \int_{-1}^{1} u d\eta = cons\tan t \tag{4.9a}$$

$$\frac{d}{dx}[\int_{-1}^{1} u^2 \delta d\eta] + \int_{-1}^{1} \{ \delta \frac{d}{dx} - (Y' + \delta'\eta) \frac{\partial}{\partial \eta} \} \frac{p}{\rho} d\eta = \int_{-1}^{1} F_x \delta d\eta \tag{4.9b}$$

$$\frac{d}{dx}\left[\int_{-1}^{1} uv\delta d\eta + \frac{p}{\rho}(x,1) - \frac{p}{\rho}(x,-1) \right] = \int_{-1}^{1} F_y \delta d\eta \tag{4.9c}$$

From (4.9a), the product of $\delta\alpha$ is a constant. For ease in derivation, we take:

$$\frac{\delta\alpha}{\pi R} \equiv K_1 \tag{4.10}$$

From (4.9b) and (4.9c), we obtain:

$$\alpha \frac{d\alpha}{dx} + \frac{1}{U_0} \frac{d}{dx}(\frac{p}{\rho}) = \frac{F_x}{U_0^2} \tag{4.11}$$

and

$$\alpha \frac{d\beta}{dx} = \frac{1}{\gamma}(\frac{F_y}{U_0^2}) \tag{4.12}$$

In the undistorted region, by integrating the equations of motion in $[1, \frac{2\pi R}{\delta} - 1]$, we thus obtain:

$$\alpha_0 \frac{d\alpha_0}{dx} + \frac{1}{U_0^2} \frac{d}{dx}(\frac{p}{\rho}) = \frac{F_{x,0}}{U_0^2} \tag{4.13}$$

and:

$$\alpha_0 \frac{d\beta_0}{dx} = \frac{1}{\bar{\gamma}} \left(\frac{F_{y,0}}{U_0^2} \right) \tag{4.14}$$

Next, with integration of the continuity equation over the whole region:

$$\delta \int_{-1}^{1} \alpha U_0 d\eta + \delta \int_{1}^{\frac{2\pi R}{\delta}-1} \alpha_0 U_0 d\eta \equiv \text{contant} \tag{4.15}$$

yields:

$$[2\delta \alpha U_0 + (2\pi R - 2\delta)(\alpha_0 U_0)] / (2\pi R U_0) \equiv K_0 \tag{4.16}$$

Using (4.10):

$$K_1 + (1 - K_1 / \alpha)\alpha_0 = K_0 \tag{4.17}$$

Rearranged as:

$$\alpha_0 = \frac{\alpha(K_0 - K_1)}{\alpha - K_1} \tag{4.18}$$

By differentiating the above equation will result in:

$$\frac{d\alpha_0}{dx} = K_2 \frac{d\alpha}{dx} \tag{4.19}$$

where:

$$K_2 = \frac{K_1(K_1 - K_0)}{(\alpha - K_1)^2} \tag{4.20}$$

The integrated result of equation of motion in the overall region is equal to the combined results in both distorted and undistorted regions. In x-direction, by combining the (4.11) and (4.13), we obtain:

$$\alpha \frac{d\alpha}{dx} - \alpha_0 \frac{d\alpha_0}{dx} = \frac{F_x - F_{x,0}}{U_0^2} \tag{4.21}$$

Substituting of (4.18) and (4.19) into (4.21), yields:

$$\alpha \frac{d\alpha}{dx} = \frac{1}{K_3} \frac{F_x - F_{x,0}}{U_0^2} \tag{4.22}$$

where:

$$K_3 = 1 + \frac{K_2(K_1 - K_0)}{\alpha - K_1} \tag{4.23}$$

From the above results, we can rearrange the five ordinary differential equations, (4.11), (4.12), (4.14), (4.19), and (4.22) as follows:

$$\alpha \frac{d\alpha}{dx} = \frac{1}{K_3}\left(\frac{F_x - F_{x,0}}{U_0^2}\right) \tag{4.24a}$$

$$\alpha \frac{d\beta}{dx} = \frac{1}{\bar{\gamma}}\left(\frac{F_y}{U_0^2}\right) \tag{4.24b}$$

$$\frac{d\alpha_0}{dx} = K_2 \frac{d\alpha}{dx} \tag{4.24c}$$

$$\alpha_0 \frac{d\beta_0}{dx} = \frac{1}{\bar{\gamma}}\left(\frac{F_{y,0}}{U_0^2}\right) \tag{4.24d}$$

$$\frac{d}{dx}\left(\frac{p}{\rho}\right) = U_0^2\left(\frac{F_x}{U_0^2} - \alpha \frac{d\alpha}{dx}\right) \tag{4.24e}$$

The above integral equations include five variables. They are two distorted velocity coefficients $\alpha(x)$ and $\beta(x)$, two undistorted velocity coefficients $\alpha_0(x)$ and $\beta_0(x)$, and one static pressure (p/ρ). These integral equations can describe the development of both distorted and undistorted regions, as well as (and most important) the progression of pressure in the compressor. The relative magnitude of vertical extension of distorted region is:

$$\xi(x) = \frac{\delta(x)}{\pi R} = K_1/\alpha(x) \tag{4.25}$$

With calculation of the four velocity coefficients in solving the integral equations, the size of distorted region, $\xi(x)$, can then be computed using (4.25).

There are two significant improvements between the current integral equation and the previous attempt. One of them is the force expression. In the previous papers, for simplicity, the lift and drag coefficients were assumed as constants and then an artificial term including the angle of attack was inserted to correct the error induced by the assumption. This simplification resulted in the vanishing of lift and drag forces simultaneously at some flow angles, which is in a real flow unrealistic (non-physical). We thus employ actual airfoil characteristics in expressing the force components without any significant assumption, which should produce flow with better physics.

The second contribution is in the derivation of integral equation. Kim et al. [6] used an equation with conservative form in distorted region but non-conservative form in undistorted region. We derive the equations in strong conservative form for both regions. This permits different results with the changes of integral equation,

(4.24d), the inclusion of variation in pressure-density ratio, (4.24e), and the associated parametric expression, (4.23). These equations are different from the previous effort.

4.3 Results and Discussion

The integral equation has been coded and solved in Chap. 1, and the results were compared with that of Kim et al. [6]. Based on these research work, the current integral equation is solved by using fourth order Runge-Kutta method with the following initial conditions: $\lambda_r = \lambda_s = 0.5\lambda$, $\alpha_0(0) = \beta_0(0) = 1.0$, and $\alpha(0) = \beta(0)$. From the experimental test cases [14], the rotor blade speed is $\omega R = 36.6\,m/s$, $\sigma = 2$, and the airfoil blade section is NACA 65-series.

4.3.1 Lift and Drag Coefficients

The most significant part in the current effort is the development of the application with airfoil characteristics.

In the first step, we release/avoid the assumption of constant lift and drag coefficients, as well as the need of correcting terms in force components. Then we collated the necessary experimental data of both lift and drag coefficients. Finally, by curve fitting procedure, two sets of data are summarized as two cure formulas suitable to be employed as the expression of force components.

The experimental data of NACA 65-series airfoil [16] indicates that the lift coefficient has a linear relationship with the angle of attack in a normal range: $[-8°, 10°]$. While the drag coefficient is the second or higher power of the lift coefficient, and there is a much smaller drag coefficient with a small angle of attack: $[-1°, 2°]$. For example, a NACA 65-209 wing section with a lower Reynolds number: $Re = 3.0 \times 10^6$, the data can be described as:

$$C_l = 0.1062\bar{\alpha} + 0.15 \quad (-8° \leq \bar{\alpha} \leq 10°) \tag{4.26a}$$

$$C_d = 0.6518 \times 10^{-2} - 0.96313 \times 10^{-3} C_l + 0.79495 \times 10^{-2} C_l^2 - \qquad (\bar{\alpha} < -1° \ \& \ \bar{\alpha} > 2°)$$
$$0.60949 \times 10^{-2} C_l^3 - 0.20416 \times 10^{-3} C_l^4 + 0.44918 \times 10^{-2} C_l^5 \tag{4.26b}$$

$$C_d = 0.0043 \quad (-1° \leq \bar{\alpha} \leq 2°) \tag{4.26c}$$

Here, $\bar{\alpha}$ is the wing section angle of attack. In the present case, (4.26a) is obtained by linear fitting, and the (4.26b) is calculated by Chebyshev curve fitting. Using (4.26) in (4.8), the force components can thus be obtained.

4.3.2 Inlet Distorted Velocity Coefficient

According to the Chap. 1, the inlet distorted velocity coefficient and incident angle are the essential parameters affecting the inlet distortion propagation. Firstly, the effect of variation in inlet distorted velocity coefficient, $\alpha(0)$, is analyzed here to show what role does it play in the current novel integral method.

To facilitate in discussion of distortion quantitatively, a distortion level is defined as:

$$\Gamma(x) = 1 - \frac{\alpha(x)}{\alpha_0(x)} \qquad (4.27)$$

The smaller $\alpha(0)$ means higher inlet distortion level. It is obvious that the definition of distortion level in representing the relative distortion is more intuitive than using the distorted velocity coefficient. For example, the case with $\alpha_0(0) = 1.0$ and $\alpha(0) = 0.1$, the distortion level at inlet is $\Gamma(0) = 0.9$. This is a severe distortion case with a high initial distortion level. While $\alpha(0) = 1.0$ will result in a $\Gamma(0) = 0.0$, and hence zero distortion.

Figure 4.4 shows that the previous work (in Chap. 1) is in good agreement with that of Kim et al. [6], which indicates that the propagation of inlet distortion with a bigger inlet distortion level will grow and vice-versa. However, the results using the present novel integral method suggest a different conclusion. From Fig. 4.5, the novel method provides a more serious propagation of inlet distortion. On the

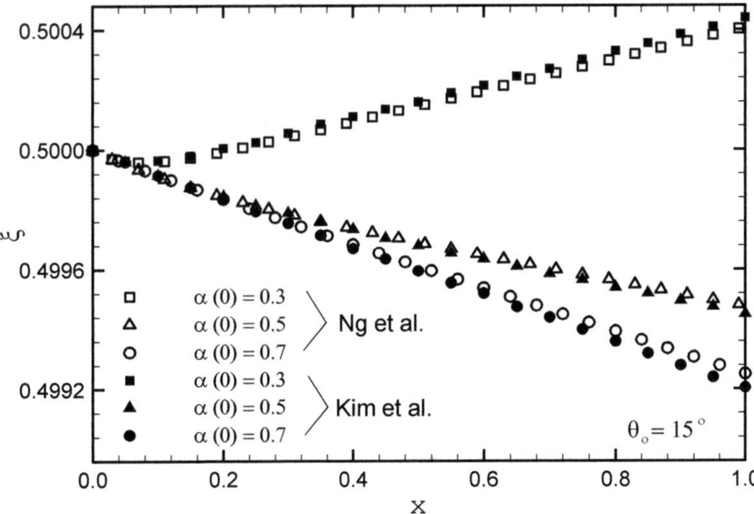

Fig. 4.4. A comparison of distorted flow propagation between the results of Ng et al.[11] and that of Kim et al. [6]

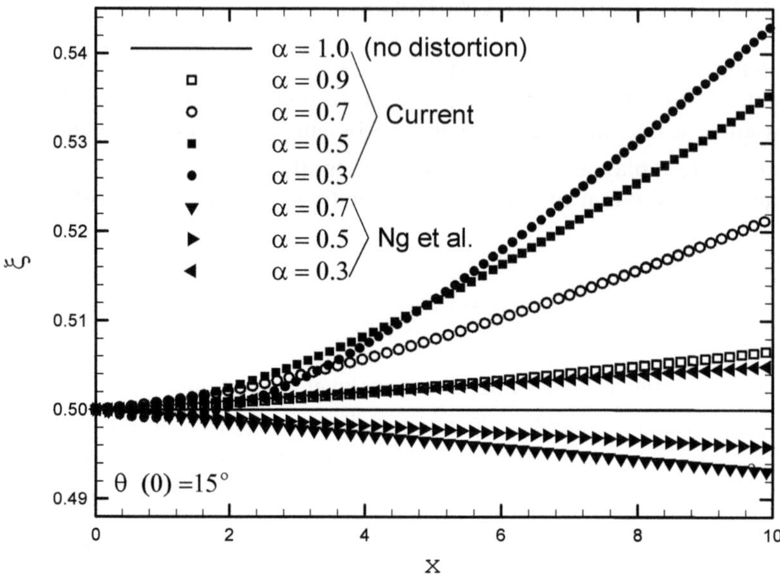

Fig. 4.5. The inlet distortion propagates along axial direction with different inlet distortion level

other hand, unlike the cases in Fig. 4.4, the present results indicate that for any inlet distortion level, the size of distorted region will grow along x-direction. In other word, using a force with simplified assumption, the integral method would underestimate the propagation of inlet distortion.

Figure 4.5 indicates that a distorted region size will increase with an increasing of distortion. Higher inlet distortion level (or smaller inlet distorted velocity coefficient) results in a more severe propagation of distortion.

4.3.3 Inlet Incident Angle

To study on extreme case, a higher inlet distortion level ($\Gamma(0) = 0.9$, or $\alpha(0) = 0.1$) is fixed during the analysis for variation in distortion with different inlet incident angles.

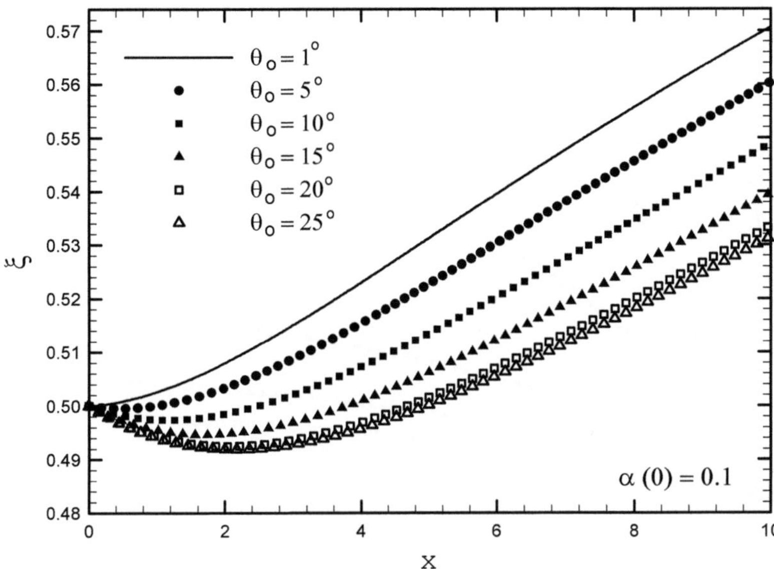

Fig. 4.6. The inlet distortion propagates along axial direction at smaller inlet incident angles, $\theta_0 \leq 25°$

The calculation shows that the inlet distortion will grow for a multistage compressor in any inlet incident angle. However, this growing magnitude is not a monotonous function of inlet incident angle only. The increment of distorted region size at outlet, $\xi(x = 10)$, will decrease with the increasing of inlet incident angle before about $\theta_0 = 25°$, and then will increase with the increasing of incident angle. Therefore, the results are presented in two figures: Fig. 4.6 and Fig. 4.8.

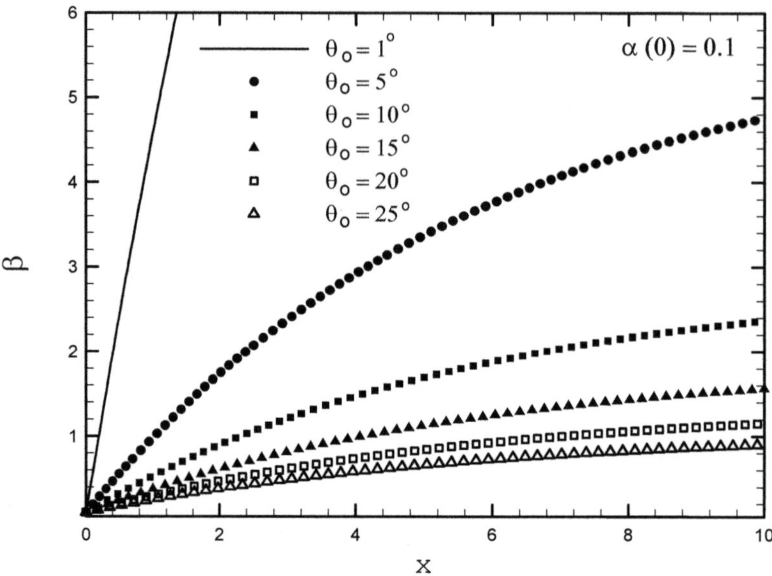

Fig. 4.7. The vertical distorted velocity coefficient propagates along axial direction at smaller inlet incident angles, $\theta_0 \leq 25°$

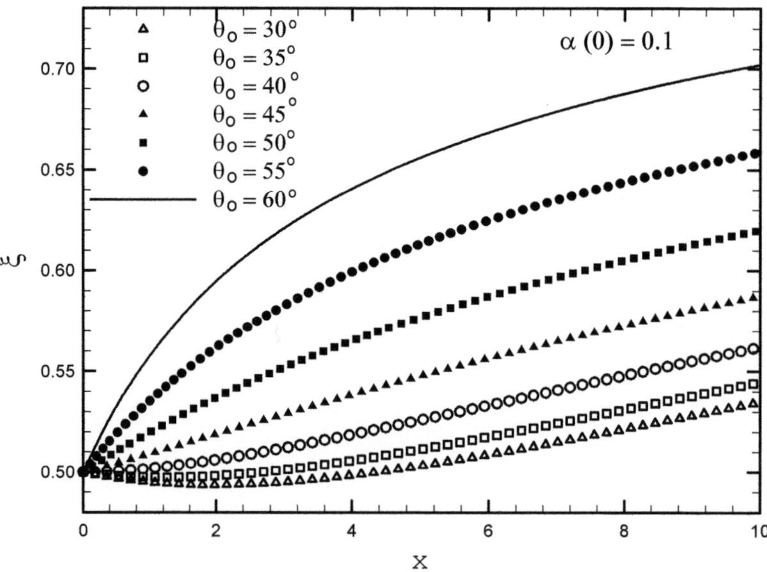

Fig. 4.8. The inlet distortion propagates along axial direction at higher inlet incident angles, $\theta_0 \geq 25°$

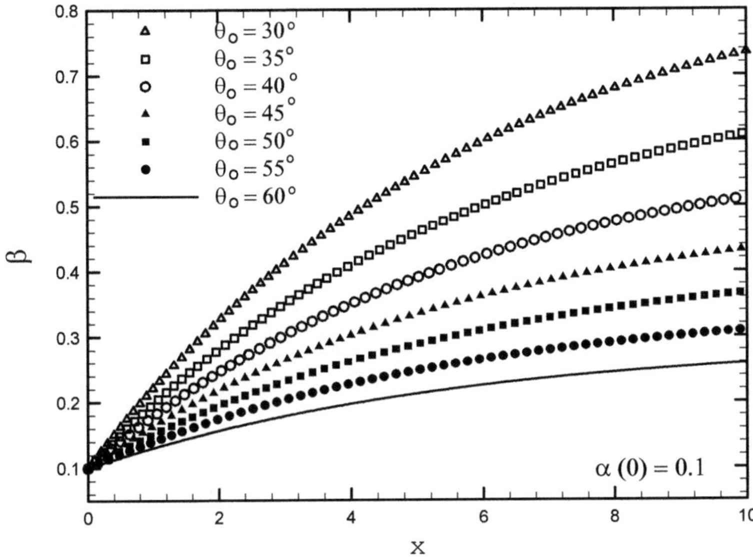

Fig. 4.9. The vertical distorted velocity coefficient propagates along axial direction at higher inlet incident angles, $\theta_0 \geq 25°$

In Fig. 4.6, smaller inlet incident angle induces a larger propagation of inlet distortion. Because a small inlet incident angle induces a large vertical flow in distorted region as shown in Fig. 4.7, thus induces a small axial distorted velocity coefficient from (4.24b), and then a large size of distorted region from (4.25). On the contrary, when the inlet incident angle grows to a large value, $\theta_0 = 25° \sim 30°$ in the current case, the increment of distorted region size at outlet will increase with the increasing of the inlet incident angle as shown in Fig. 4.8. This is because with a larger inlet incident angle, the vertical flow in distorted region tends to decrease (Figure 4.9).

4.3.4 Propagation of Distortion Level

The inlet distortion varying along axial direction with different inlet velocity coefficients or inlet flow angles has been investigated. However, what would be observed from the viewpoint at outlet for a ten-stage compressor with different inlet velocity coefficients, inlet flow angles or inlet distorted region sizes? Figure 4.10 and Fig. 4.11 indicate that the outlet size of distorted region is larger for a case with higher inlet distortion level regardless of what the inlet size of distorted region is. On the other hand, for a case with higher inlet distortion level, the radius of curvature of outlet size of distorted region tends to be increased whatever the inlet size of distorted region is.

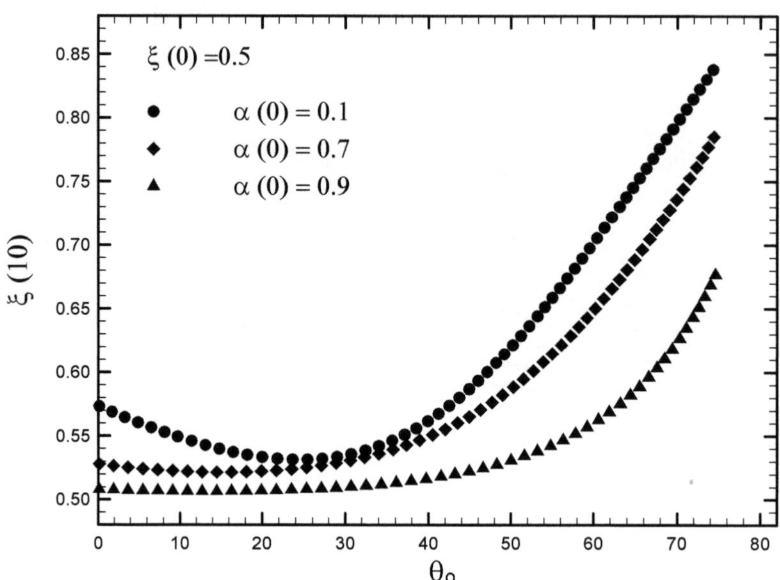

Fig. 4.10. The predicted outlet size of distorted region vs. θ_0 with higher inlet size of distorted region of 0.5

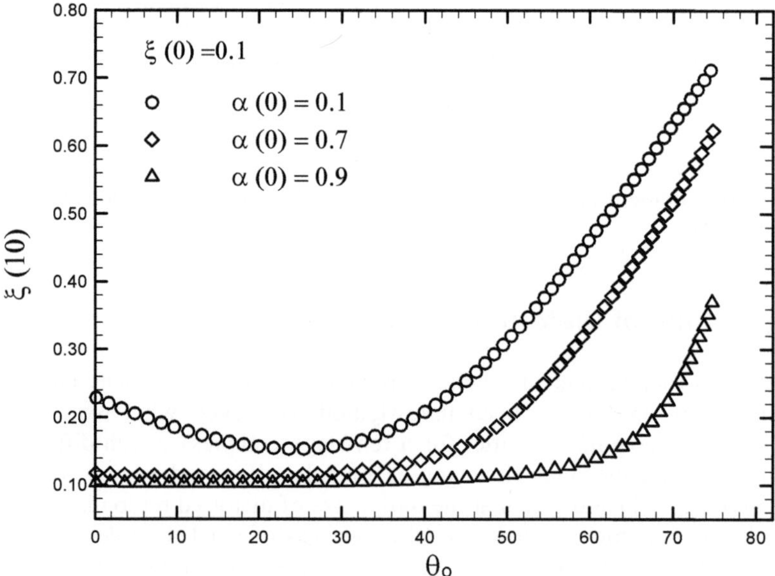

Fig. 4.11. The predicted outlet size of distorted region vs. θ_0 with smaller inlet size of distorted region of 0.1

When inlet flow angle is very small, the outlet size of distorted region will decrease with the decreasing of inlet distortion level. With the increase of inlet flow angle, the peak point of outlet size of distorted region will move forward along $\alpha(0)$ axes (Figure 4.12 and Fig. 4.13). In other words, the peak point of $\xi(10)$ corresponds to an increased value of $\alpha(0)$ at a higher inlet flow angle.

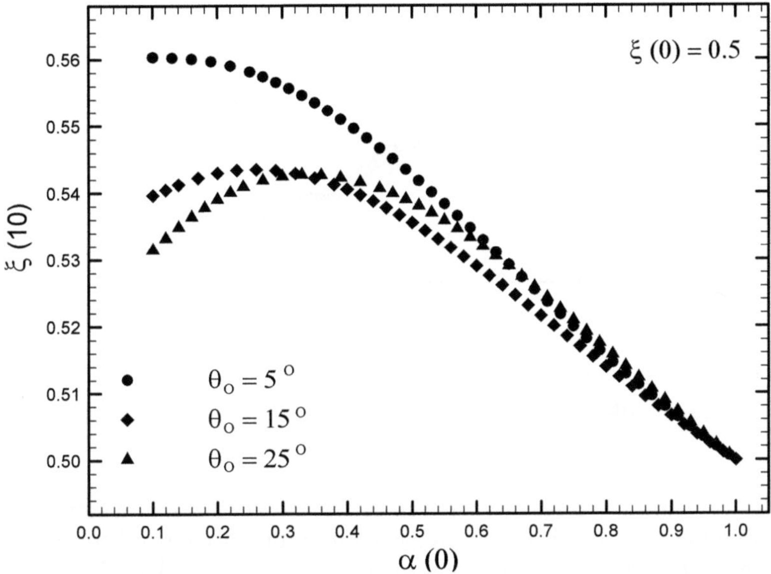

Fig. 4.12. The predicted outlet size of distorted region vs. $\alpha(0)$ with higher inlet size of distorted region of 0.5

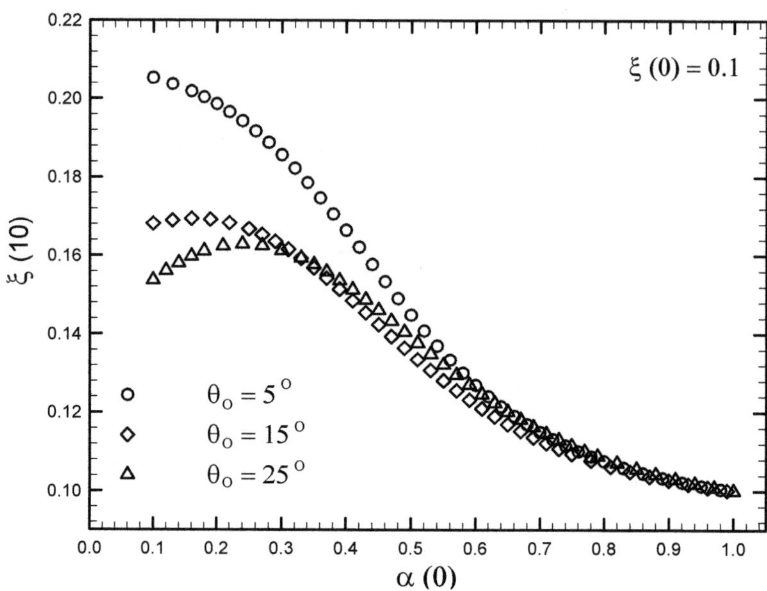

Fig. 4.13. The predicted outlet size of distorted region vs. $\alpha(0)$ with smaller inlet size of distorted region of 0.1

To ease in comparing the results between different inlet sizes of distorted region, we define the level of distortion propagation by the difference between the sizes of outlet and inlet distorted regions such as $\left[\xi(10) - \xi(0)\right]$ in the current case. With this definition, we can arrange the results with different inlet sizes of distorted region in a single plot as shown in Fig. 4.14 and Fig. 4.15. Both figures illustrate that for a higher inlet flow angle, more severe distortion propagation occurs with a larger inlet size of distorted region. On the contrary, for a lower inlet flow angle with $\theta_0 \leq 25°$, a higher level of distortion propagation occurs with a smaller inlet size of distorted region $\xi(0)$.

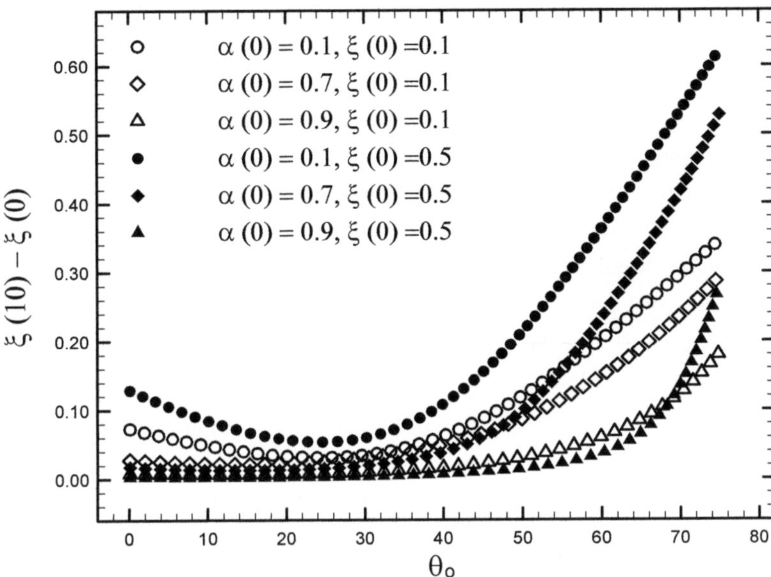

Fig. 4.14. The level of distortion propagation versus inlet flow angle

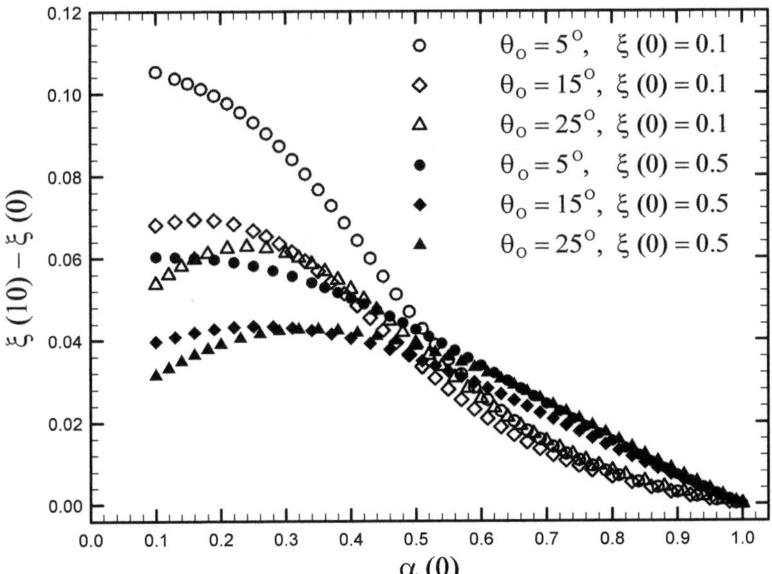

Fig. 4.15. The level of distortion propagation versus inlet distorted velocity coefficient

4.3.5 Compressor Characteristics

The total pressure ratio and the static pressure rise of compressor are investigated to study the effects of inlet parameters on the compressor performance and characteristics.

Figure 4.16 indicates that a smaller inlet flow angle causes a higher total pressure ratio, and a smaller inlet distorted velocity coefficient $\alpha(0)$, or a higher inlet distortion level $\Gamma(0)$ induces a higher total pressure ratio. However, the inlet size of distorted region has no obvious effect on the total pressure ratio.

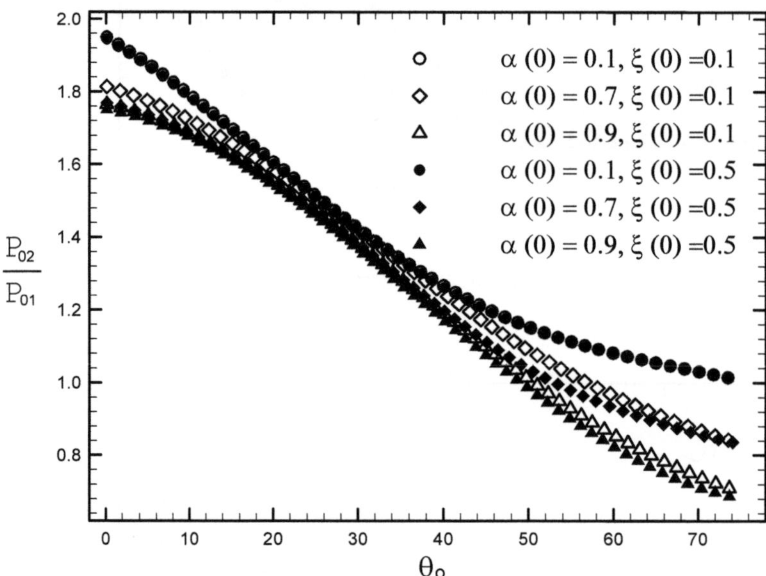

Fig. 4.16. The computed compressor total pressure ratio versus inlet flow angle

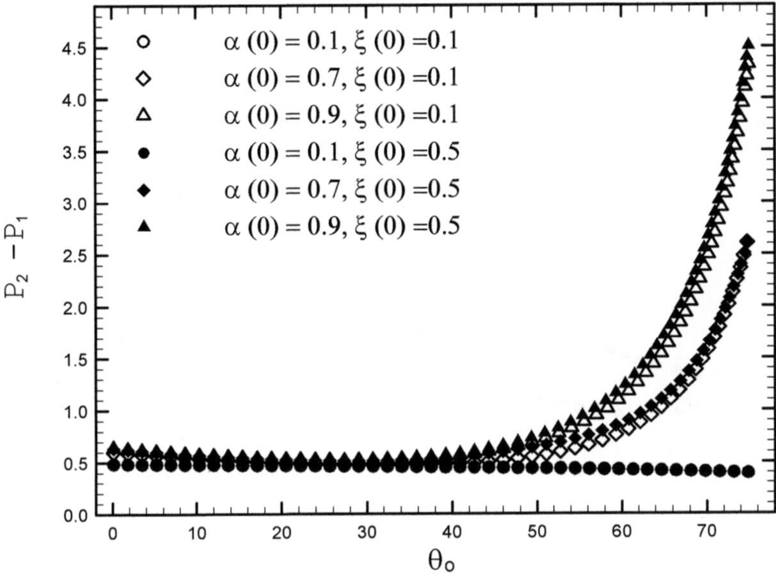

Fig. 4.17. The computed compressor static pressure rise versus inlet flow angle

In the small and medium inlet flow angles, the static pressure rises are almost the same with a constant value as shown in Fig. 4.17. The static pressure rise does not change whatever the inlet distorted velocity coefficient, $\alpha(0)$, and the inlet size of distorted region, $\xi(0)$, are. However, with a larger inlet flow angle ($\theta_0 > 45°$), the static pressure rises sharply with a higher inlet distorted velocity coefficient.

By keeping eyes on the smaller inlet flow angles with $\theta_0 \leq 25°$ and changing $\alpha(0)$ from 0 to 1.0, we can obtain the compressor characteristics as summerised in Fig. 4.18 and Fig. 4.19. The larger inlet size of distorted region, $\xi(0)$, produces a larger change in mass flow rate. Meanwhile, a smaller flow angle produces a higher pressure rise. This is easy to understand. Because $\sigma = \dfrac{\omega R}{V_0}$, and we set σ and ωR as constants, thus the vertical reference velocity V_0 is a constant too. From the (4.3), $\tan\theta_0 = V_0/U_0$, a smaller θ_0 will produce a larger U_0 for a constant value of V_0, thus a higher pressure rise.

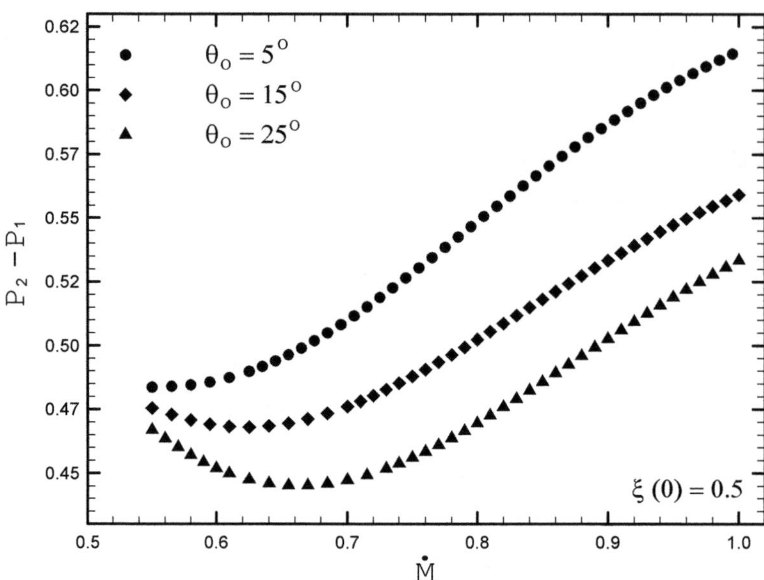

Fig. 4.18. The simulated compressor characteristics at higher inlet size of distorted region of 0.5

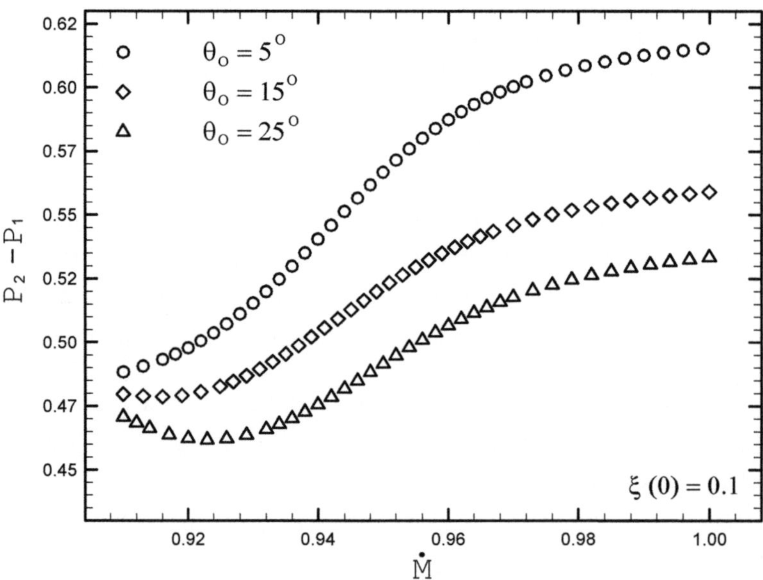

Fig. 4.19. The simulated compressor characteristics at smaller inlet size of distorted region of 0.1

4.3.6 Argument for Airfoil Characteristics

As mentioned in Sect. 4.3.1, we propose the airfoil lift and drag coefficients to be described as the functions of the angle of attack according to the wing section theory and experimental data:

$$C_l = f_1(\bar{\alpha}) \tag{4.28a}$$

$$C_d = f_2(C_l) \tag{4.28b}$$

Here, the angles of attack for rotor and stator are described as following in the distorted region (Figure 4.3),

$$\bar{\alpha}_r = tan^{-1}(\frac{\sigma - \beta}{\alpha}\bar{\gamma}) \tag{4.29a}$$

$$\bar{\alpha}_s = tan^{-1}(\frac{\beta}{\alpha}\bar{\gamma}) \tag{4.29b}$$

The angles of attack in the undistorted region can also be obtained with expression of undistorted velocity coefficients α_0 and β_0.

From (4.24a) and (4.25), we have:

$$\frac{d\xi}{dx} = \frac{1}{\dfrac{K_1^2}{\xi^3} + \dfrac{(K_1 - K_0)}{(1-\xi)^3}} \cdot \frac{F_{x,0} - F_x}{U_0^2} \tag{4.30}$$

Because the first term of the right hand side is positive, $\dfrac{K_1^2}{\xi^3} + \dfrac{(K_1 - K_0)}{(1-\xi)^3} > 0$, thus

the change of distorted region size, $\dfrac{d\xi}{dx}$, or the level of distortion propagation of a

ten-stage compressor, $\xi(10) - \xi(0)$, would depend on the airfoil forces in both the distorted and undistorted regions. Therefore, it is vital to apply the appropriate/associated airfoil characteristics in the prediction of inlet distortion propagation, and the calculated results would be more realistic.

4.4 Concluding Remarks

The original integral method for investigating a compressor inlet distortion has been developed further, and a novel method is derived by utilizing the more appropriate and practical airfoil characteristics. The simulation indicates that results calculated using the previous integral method underestimated the level of distortion propagation as compared to the current novel integral method. The investigation of inlet parameters is also proceeded and the results suggest that the inlet distortion level and the inlet flow angle have noticeable effects upon the downstream flow features, especially in the level of distortion propagation.

The present effort suggests that this simple method may be flourishing in the problems of strongly distorted flow and propagating stall in axial compressor. It is believe that this approach could be developed further by using a more realistic and flexible velocity and pressure profiles.

References

[1]   Chen J.P., and Briley W.R., 2001, A parallel flow solver for unsteady multiple blade row turbomachinery simulations. ASME paper 2001-GT-0348.

[2]   Cumpsty N.A. and Greitzer E.M., 1982, A simple model for compressor stall cell propagation. Transactions of the ASME, Journal of Engineering for Power, **104**: 170-176.

[3]   Emmons H.W., Pearson C.E. and Grant H.P., 1955, Compressor surge and stall propagation. Transactions of the ASME, **77**: 455-469.

[4]   Greitzer E.M., 1980, Review: axial compressor stall phenomena. ASME Journal of Fluids Engineering, **102**: 134-151.

[5]   Jonnavithula S., Thangam S. and Sisto F., 1990, Computational and experimental study of stall propagation in axial compressors. AIAA Journal, **11**: 1945-1952.

[6]   Kim J.H., Marble F.E., and Kim C-J., 1996, Distorted inlet flow propagation in axial compressors. In Proceedings of the 6th International Symposium on Transport Phenomena and Dynamics of Rotating Machinery, Hawaii, **2**: 123-130.

[7]   Longley J.P. and Hynes T.P., 1990, Stability of flow through multistage axial compressors. Transactions of the ASME, Journal of turbomachinery, **112**: 126-132.

[8]   Marble F.E., 1955, Propagation of stall in a compressor blade row. J. Aero. Sci., **22(8)**: 541-554.

[9]   Mazzawy R.S., 1977, Multiple segment parallel compressor model for circumferential flow distortion, ASME Journal of Engineering for Power, **99(4)**: 288-296.

[10]   Mikolajczak A.A. and Pfeffer A.M., 1974, Method to increase engine stability and tolerance to distortion, AGARD 72.

[11]   Ng E.Y-K., Liu N., Lim H.N. and Tan T.L., 2002, Study on The Distorted Inlet Flow Propagation In Axial Compressor Using An Integral Method, Computational Mechanics, **30(1):** 1-11.

[12]   Plourde G.A. and Stenning A.H., 1967, Attenuation of circumferential inlet distortion in multistage axial compressors, AIAA Paper 67-GT-415.

[13]   Reid C., 1969, The response of axial flow compressors to intake flow distortion. In Proceeding of International Gas Turbine and Aeroengine Congress and Exhibition, ASME Paper 69-GT-29.

[14]   Serovy G.K. and Dring R.P., 1990, Experimental test cases for compressors. In "Test cases for computation of internal flows in aero engine components", edited by Fottner L, AGARD-AR-**275:** 151-164.

[15]   Stenning A.H., 1980, Inlet distortion effects in axial compressors. ASME Journal of Fluids Eng., **102(3):** 7-13.

[16]   Von Doenhoff E.A., 1959, Theory of wing sections, including a summary of airfoil data. New York, Dover Publications.

[17]   Wellborn S.R., and Delaney R.A., 2001, Redesign of a 12-stage axial-flow compressor using multistage CFD. ASME paper 2001-GT-0315.

Appendix 4.A Fortran Program: Chebyshev Curve Fitting

For convenience in analyzing and comparing to a give experimental data or previous computational results, it is necessary to curve fit them. The easiest one is the polynomial fitting. In current work, we need the curve formulas from the experimental data of both lift and drag coefficients. Therefore, a curve fitting is adopted.

This code applies the Chebyshev curve fitting (as mentioned following) to find the best multinomial suitable to the given data.

Given n data of points: $((x_i, y_i), i = 1, 2, ..., n$ with $x_1 < x_2 < ... < x_n$, to find a (m-1)th order multinomial(m<n): $P(x) = a_1 + a_2 x + ... + a_m x^{m-1}$, and make sure there are smallest deviations at the given points: $\max_{1 \le i \le n} |P(x_i) - y_i| = min$.

1. The input data are

X   :   double precision one dimensional data set depositing N data
Y   :   double precision one dimensional data set depositing N data
N   :   input parameter, the number of data points
A   :   double precision one dimensional data set depositing M1=M+1 data.
M   :   input parameter, the number of multinomial item, and its order is M-1.
M1   :   input parameter, M1=M+1.

2. Requirement:

M<N and M$\le$19

3. Example:

A set of data comes from 20 point:
X_i=0.14,0.16,0.18,0.20,0.22,0.24,0.26,0.28,0.30,0.32,
     0.34,0.36,0.38,0.40,0.415,0.42,0.44,0.46,0.48,0.49
Y_i=0.855,0.845,0.837,0.832,0.830,0.830,0.832,0.837,0.845,0.855,
     0.865,0.880,0.900,0.925,0.955,0.960,0.995,1.040,1.080,1.10
To construct a Chebyshev multinomial :
$$P(x) = a_1 + a_2 x + a_3 x^2 + a_4 x^3 + a_5 x^4 + a_6 x^5$$
Therefore, N=20, M=6, M1=7.

To run the program, a result is produced as:

A(1)=1.77546069
A(2)=-16.6288937
A(3)=115.42272
A(4)=-394.29835
A(5)=657.747843
A(6)=-416.677009
HMAX=0.0030167777

Here, A(1)...A(6) are $a_1...a_6$, and HMAX is the biggest deviation.

```fortran
*****************************************************************
*                  Chebyshev Curve Fitting                     *
*                  -----        MPE, NTU, SINGAPORE            *
*****************************************************************
          DIMENSION X(20),Y(20),A(7)
          DOUBLE PRECISION X,Y,A
          DATA X/0.14,0.16,0.18,0.20,
     &    0.22,0.24,0.26,0.28,0.30,0.32,0.34,
     &    0.36,0.38,0.40,0.415,0.42,0.44,0.46,0.48,0.49/
          DATA Y/0.855,0.845,0.837,0.832,
     &    0.830,0.830,0.832,0.837,0.845,0.855,0.865,
     &    0.880,0.900,0.925,0.955,0.960,0.995,1.040,1.080,1.10/
          OPEN(UNIT=1,FILE='CURVE.DAT')
          N=20
          M=6
          M1=7
          CALL HCHIR(X,Y,N,A,M,M1)
          WRITE(1,20)(I,A(I),I=1,M)
          WRITE(1,30)A(M1)
          WRITE(*,20)(I,A(I),I=1,M)
          WRITE(*,30)A(M1)
20        FORMAT (1X,'A(',I2,')=',D16.9)
30        FORMAT (1X,'HMAX=',D15.8)
          CLOSE(1)
          END


          SUBROUTINE HCHIR(X,Y,N,A,M,M1)
          DIMENSION X(N),Y(N),A(M1),IX(20),H(20)
          DOUBLE PRECISION X,Y,A,H,HA,HH,Y1,Y2,H1,H2,D,HM
```

```
      DO 5 I=1,M1
5     A(I)=0.0
      IF(M.GE.N) M=N-1
      IF(M.GE.20) M=19
      M1=M+1
      HA=0.0
      IX(1)=1
      IX(M1)=N
      L=(N-1)/M
      J=L
      DO 10 I=2,M
          IX(I)=J+1
          J=J+L
10    CONTINUE
20    HH=1.0
      DO 30 I=1,M1
          A(I)=Y(IX(I))
          H(I)=-HH
          HH=-HH
30    CONTINUE
      DO 50 J=1,M
          II=M1
          Y2=A(II)
          H2=H(II)
          DO 40 I=J,M
              D=X(IX(II))-X(IX(M1-I))
              Y1=A(M-I+J)
              H1=H(M-I+J)
              A(II)=(Y2-Y1)/D
              H(II)=(H2-H1)/D
              II=M-I+J
              Y2=Y1
              H2=H1
40        CONTINUE
```

```
50          CONTINUE
            HH=-A(M1)/H(M1)
            DO 60 I=1,M1
60          A(I)=A(I)+H(I)*HH
            DO 80 J=1,M-1
                II=M-J
                D=X(IX(II))
                Y2=A(II)
                DO 70 K=M1-J,M
                    Y1=A(K)
                    A(II)=Y2-D*Y1
                    Y2=Y1
                    II=K
70              CONTINUE
80          CONTINUE
            HM=ABS(HH)
            IF (HM.LE.HA) THEN
                A(M1)=-HM
                RETURN
            END IF
            A(M1)=HM
            HA=HM
            IM=IX(1)
            H1=HH
            J=1
            DO 100 I=1,N
                IF (I.EQ.IX(J)) THEN
                    IF (J.LT.M1) J=J+1
                ELSE
                    H2=A(M)
                    DO 90 K=M-1,1,-1
90                  H2=H2*X(I)+A(K)
                    H2=H2-Y(I)
                    IF (ABS(H2).GT.HM) THEN
```

```fortran
                  HM=ABS(H2)
                  H1=H2
                  IM=I
               END IF
            END IF
100      CONTINUE
         IF (IM.EQ.IX(1)) RETURN
         I=1
110      IF (IM.GE.IX(I)) THEN
              I=I+1
              IF (I.LE.M1) GOTO 110
         END IF
         IF (I.GT.M1) I=M1
         IF (I.EQ.(I/2)*2) THEN
              H2=HH
         ELSE
              H2=-HH
         END IF
         IF (H1*H2.GE.0.0) THEN
              IX(I)=IM
              GOTO 20
         END IF
         IF (IM.LT.IX(1)) THEN
              DO 120 J=M,1,-1
120           IX(J+1)=IX(J)
              IX(1)=IM
              GOTO 20
         END IF
         IF (IM.GT.IX(M1)) THEN
              DO 130 J=2,M1
130           IX(J-1)=IX(J)
              IX(M1)=IM
              GOTO 20
         END IF
```

```
IX(I-1)=IM
GOTO 20
END
```

Chapter 5
Parametric Study of Greitzer's Instability Flow Model Through Compressor System Using Taguchi Method

The occurrence of stall and surge is caused by instability of the flow through the compressor system as a whole. These two phenomena often result in serious mechanical problems for the compressor. To further illustrate the application of Taguchi method and the important of parametric study, this chapter will give another case study, including a review and parametric study on the characteristics of stall and surge and their mathematical modeling.

In the current chapter, the Greitzer's B-parameter model is applied for analyzing the stall and surge characteristics. Based on this model, a total of four parameters B, G, K and L_C in the model are highlighted in order to establish the influence of each parameter on the system. Parameters B, G and K are all dimensionless constants while L_C is the effective length of the compressor duct.

First of all, the governing equations of stall and surge behavior are solved numerically using fourth-order Runge-Kutta method. The Taguchi method is then used to analyze the results generated to obtain the extent of effects of the parameters on the system by varying the parameters in a series of combinations. Finally, a thorough analysis is carried out on the results generated from the Taguchi method and the graphs. It is found that parameter K is the deciding factor in the position of the cross point whereas parameter B is the most important one in changing the length of time needed for the compressor to reach its steady state.

5.1 Introduction

There are two types of instabilities that could be encountered in axial compressor systems. They are rotating stall and surge respectively. Both types of instability have damaging consequences to the compressor.

Rotating stall causes the compressor to operate with extremely low frequencies, which results in excessive high internal temperature that has an adverse effect on blade life. Surge can cause severe problems such as excessive built-up of pressure at the inlet and cyclic loading on compressor mounting. If the instability occurs, the blades might fail to produce required loading, and the engine might occur catastrophic damage. To understand and avoid compressor stall and surge, it is important to know the characteristics of these two phenomena.

The surge and rotating stall are serious problems for the axial compressor system. As their consequences are different, it is useful to be able to predict which of these problems will occur for a given situation.

To know which will occur in a particular situation, the detail measurement and theoretical analysis have been taken during the stall inception processes ([1], [8] and [12]). The studies for the stall inception confirmed that surge in axial compressor is initiated by rotating stall [2]. Thus, it should be possible to inhibit the onset of surge by suppressing the occurrence of rotating stall.

In this area, the active control techniques have achieved a great success in suppressing the fluid dynamic instabilities that lead to stall and surge ([7] and [2]). The study of active control techniques, on the contrary, is helpful at understanding the stall inception process, and the detail flow condition in compressors during rotating stall or surge ([2], [3], [4], [5] and [7]).

As the works of understanding the flow condition and compressor performance, the effective analysis and modeling can be traced back to the work of Greitzer in 1976 [5], then Moore and Greitzer [9], and Greitzer and Moore [6] in 1986, and others. Among these works, the Greitzer's B-parameter model [5] is well known and effective one for analyzing the stall and surge characteristics. It is meaningful to study and further understand this model using current numerical method and computation facility.

In the current chapter, based on the Greitzer's B-parameter model, the authors solve numerically the governing equations of stall and surge behavior using fourth-order Runge-Kutta method. Then using Taguchi method ([10] and [11]) we analyze the results generated to obtain the extent of effects of the parameters on the system by varying the parameters in a series of combinations. Finally, a thorough analysis is done and the conclusion about the deciding factor for compressor characteristics is worked out.

5.2 Mathematical Model

Greitzer [5] set up a mathematical model to analyze the characteristics of stall and surge. This model maps out not only the compressor but also the whole compressor system and may be used to predict whether a stall or a surge will occur at stall limit.

As shown in Fig. 5.1, a compressor system normally consists of the following components: a compressor, an inlet annular duct, a plenum and a throttle in an exit duct whose diameter is much smaller than that of the plenum.

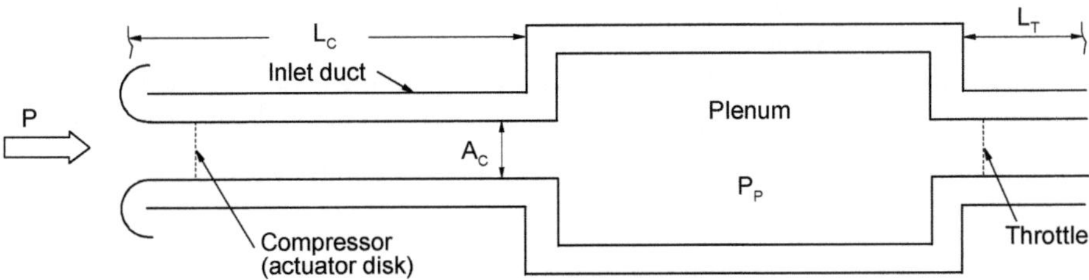

Fig. 5.1. Simplified diagram of compression system

Apply the equations of motions to the these components of the compressor system, we can obtain

$$\frac{d\dot{\tilde{m}}_C}{d\tilde{t}} = B(\tilde{C} - \Delta\tilde{P})$$
(5.1a)

$$\frac{d\dot{\tilde{m}}_T}{d\tilde{t}} = \frac{B}{G}(\Delta\tilde{P} - K\dot{\tilde{m}}^2)$$
(5.1b)

$$\frac{d\Delta\tilde{P}}{d\tilde{t}} = \frac{1}{B}(\dot{\tilde{m}}_C - \dot{\tilde{m}}_T)$$
(5.1c)

$$\frac{d\tilde{C}}{d\tilde{t}} = \frac{1}{\tilde{\tau}}(\tilde{C}_{SS} - \tilde{C})$$
(5.1d)

Three non-dimensional parameters are also included in these equations. They are

$$B = \frac{U}{2a}\sqrt{\frac{V_P}{A_C L_C}} \; ; \qquad G = \frac{L_T A_C}{L_C A_T} \; ; \qquad K = \frac{A_C^{\,2}}{A_T^{\,2}} \cdot$$

Here $\tilde{\tau}$ is the non-dimensional time lag; ω is the Helmholtz frequency; $\Delta P = P_P - P$ is the pressure difference across the duct; and C is the pressure rise across the compressor. $\dot{m}$ is the mass flow rate. L and A are the length and area. τ and C_{SS} are the compressor flow field time constant and the steady-state measured compressor curve respectively. N is the time for some number of rotor revolutions. The subscripts C, P and T are the compressor duct, plenum and throttle duct respectively. The subscript O is initial non-dimensionalized value at surge point (or stall point). The top cap ~ is the non-dimensionalized quantity. The non-dimensional parameters are $\rho U A_C$, the pressure difference variable $0.5\rho U^2$ and the time variable $1/\omega$.

Equation (1) is the non-dimensionalized 1-D governing equation for compressor system at transient state, which is the so-called Greitzer's B-parameter model. The dimensionless B-parameter represents geometrical and physical parameters of system. Low values of B favor rotating stall characteristic and high values of B favor surge like oscillation whereas similar values of B give similar flow behavior. For high values of B (implying very large plenum volume) the plenum pressure remains high as the compressor starts to stall. The resulting large pressure difference across the compressor duct is enough to rapidly decelerate the flow. For low values of B (small volumes), the plenum pressure drops quickly compared with the time it takes to reverse the flow in the compressor ducting. There is then an insufficient pressure difference tending to drive the flow backward through the compressor and the operating point settles down on the rotating stall characteristic. It is indeed a critical B^* (≈ 0.7) which separates systems that exhibit surge and those that exhibit rotating stall (i.e. transient behavior after initial instability depends on B). In brief, surge will occur whenever the release of the stored energy of the high pressure fluid is enough to overcome the inertia of the fluid in the compressor ducting (i.e. we can view B parameter as a ratio of stored energy to work necessary to overcome inertia forces). Parameter G is a volume ratio (of cross product of effective length of throttle and compressor through flow area to effective length of compressor ducting and throttle through flow area, G can be interpreted as an pointer on flow continuity). Parameter K is the through flow area ratio of compressor to throttle (high K values imply throttling capacity is small).

To sum up, Greitzer's model is instructive since it stresses the need to correctly model the capacity of a system to store energy, and the need to correctly account for the inertia of the fluid in the ducting. This is vital in the sense that whether or not a system will surge is not a property of the compressor, but of the compressor and the system of which it is part. Three parameters, B, G and K, are used in this model. Together with the characteristic length L_C, these four parameters will be investigated in a series of combinations to analyze the effects of each parameter on the whole compressor system using the Taguchi method.

5.3 Numerical Methods

In the current effort, the characteristics of stall and surge of a particular compressor system is analyzed using the numerical method. Thereafter, Taguchi method is applied to perform a parametric optimization for compressor identification, and the fourth order Runge-Kutta method is used to solve the governing equation (1).

An empirical compressor pressure rise, $\widetilde{C}_{SS}$, is formulated according to Greitzer [5]. The initial conditions, at $t=0$, are taken as:

1) $\Delta\widetilde{P} = \widetilde{C} = \widetilde{C}_O$

2) $\dot{\widetilde{m}}_C = \dot{\widetilde{m}}_T = \dot{\widetilde{m}}_O$

where $\widetilde{C}_O$ and $\widetilde{m}_O$ are the initial non-dimensionalized compressor pressure rise and mass flow.

Taguchi method ([10] and [11]) is a very efficient tool for analyzing and developing high quality products at a low cost. Traditionally, one tends to change only one variable of an experiment at a time. The strength of the Taguchi technique is that one can change many variables at the same time and still retain control of the experiment.

In Taguchi method, the experiments shall be conducted by using a series of combinations from the different factors and levels of the input design variables. This may not be always feasible in a real experiment due to many practical difficulties in conducting the experiment such as a very high temperature/pressure study, hazardous conditions, cost, time etc. However, the numerical experiments have the great advantage in the cases of combining Taguchi method and numerical technique.

In Taguchi method, the grouping of the iterations from the various parameters of the input design variables is called an orthogonal array. Each parameter is referred as a factor and each different value for each factor is termed as a level. The number of combinations is obtained using the following formula:

Number of combinations = (Number of levels) $^{\text{Number of factors}}$

Hence for tabulations involving 3 factors and 3 levels as done in this work, number of combinations $=3^3=27$.

In the present parametric research, we solve the governing equation (5.1) by using orthogonal arrays, $L_{27}(3^{13})$, to identify the factor/variable that has the most influence on the rotating stall/surge so as to minimize it in the actual operation of the axial compressor.

From Greitzer [5], the most critical parameter will be B with G having only minor effect. Therefore G is kept constant at the value of 0.36. However in the present work, G will be analyzed systematically using Taguchi method to verify the above assumption. Hence the values of G will be set at three levels with 0.16, 0.36 and 0.56 respectively.

In the Greitzer's paper [5], surge will occur when the critical value of B is approximately 0.7. Hence in the current study, the value of B used will be kept below 0.7 to avoid the occurrence of surge with the three values being 0.50, 0.55 and 0.60 respectively.

Table 5.1. Case 1: Factors and levels for parameters B, G and K

Factors	Levels		
	1	2	3
B	0.50	0.55	0.60
G	0.16	0.36	0.56
K	5.32	5.52	5.72

Table 5.2. Case 2: Factors and levels for parameters B, G and L_C

Factors	Levels		
	1	2	3
B	0.50	0.55	0.60
G	0.16	0.36	0.56
L_C	1.36	1.46	1.56

As for the other two parameters, K and L_C, their effects on the system are not mentioned in previous paper. Therefore in the current work, these effects on the system will also be analyzed to complete the whole spectrum of parametric study. The two parameters will be categorized into two separate cases with parameters B and G as listed in Table 5.1 and Table 5.2.

Upon executing the program, the results are compiled and included/tabulated in Table 5.3 and Table 5.4.

As shown in Table 5.3 and Table 5.4, there are columns for values of *Ave X* and *Ave Y*, which are calculated from the average of the X and Y coordinates of the cross point *(X, Y)* of the compressor curve when it stabilizes to a stationary point.

The values for *Ave Diff M* and *Ave Diff P* are obtained similarly. First of all the maximum and minimum values of the instantaneous pressure rise P and the compressor axial velocity parameter M are recorded in the process of prediction. Then the difference between these two ranges of values, *Diff M* and *Diff P* are calculated and averaged subsequently.

Table 5.3. Case 1: Results for effects of parameters B, G and K on the compressor system

	Ave X	Ave Y	Ave Diff M	Ave Diff P	R(X)%	R(Y)%	R(Diff M)%	R(Diff P)%
	0.414529	0.951680	0.042722	0.054561				
B	0.415944	0.954201	0.062836	0.093696	0.142237	0.252039	5.360667	10.329344
	0.415951	0.954045	0.096328	0.157855				
	0.414513	0.951581	0.074464	0.114550				
G	0.415952	0.954147	0.065120	0.098131	0.144583	0.261614	1.216133	2.111956
	0.415959	0.954197	0.062302	0.093431				
	0.426744	0.972833	0.060423	0.095681				
K	0.415472	0.952675	0.065525	0.099169	2.253574	3.841621	1.551422	1.558044
	0.404208	0.934417	0.075937	0.111262				

Table 5.4. Case 2: Results for effects of parameters B, G and L_C on the compressor system

	Ave X	Ave Y	Ave Diff M	Ave Diff P	R(X)%	R(Y)%	R(Diff M)%	R(Diff P)%
	0.415453	0.952757	0.037022	0.045106				
B	0.415459	0.952761	0.063036	0.094409	0.002900	0.034756	5.959733	11.358556
	0.415482	0.952413	0.096619	0.158691				
	0.415422	0.952525	0.068912	0.105351				
G	0.415479	0.952649	0.065314	0.098809	0.007220	0.023289	0.646244	1.130422
	0.415494	0.952757	0.062450	0.094047				
	0.415468	0.952743	0.057156	0.079110				
L_C	0.415472	0.952675	0.065525	0.099164	0.001674	0.022910	1.683922	4.082233
	0.415455	0.952514	0.073996	0.119932				

The averages of the values X, Y, *Diff M* and *Diff P* are calculated according to their levels. This system of calculation will apply to all the three levels of the three parameters for the two cases.

Last but not least, the final results indicating the effects of the parameters on the compressor system are tabulated. Referring to the last four columns of Table 5.3 and Table 5.4 which are represented by *R(X)%*, *R(Y)%*, *R(Diff M)%*, and *R(Diff P)%*, they are the percentage values signifying the importance of the parameters. The higher the percentage value, the greater the effect of the parameter has on the system.

Figures 5.2, 5.3, 5.4, 5.5, 5.6, 5.7, 5.8, and 5.9 are plotted from the data files generated upon the execution of the program. The graphs drawn will be divided into two categories. In the first category, the graphs (Figure 5.2- Figure 5.5) will focus on the changes in the cross point *(X, Y)*. In the second category, the changes in the "diameters of oscillations" (*Diff M* and *Diff P*) will be investigated (Figures 5.6, 5.7, 5.8 and 5.9). In each figure, only one parameter is varied accordingly to their three levels, and the other parameters will be set at its original value as proposed in previous study, *B=0.60, G=0.36, K=5.52* and *L_C=1.46*.

5.4 Results and Discussion

The analysis of the influences of parameters B, G, K and L_C will be based on the results obtained upon performing the Taguchi method on these four parameters. The results can be found in the last four columns of Table 5.3 and Table 5.4 respectively. As seen from the values obtained, they are presented in percentage values. This implies that the higher the percentage value, the higher the influence of the parameter.

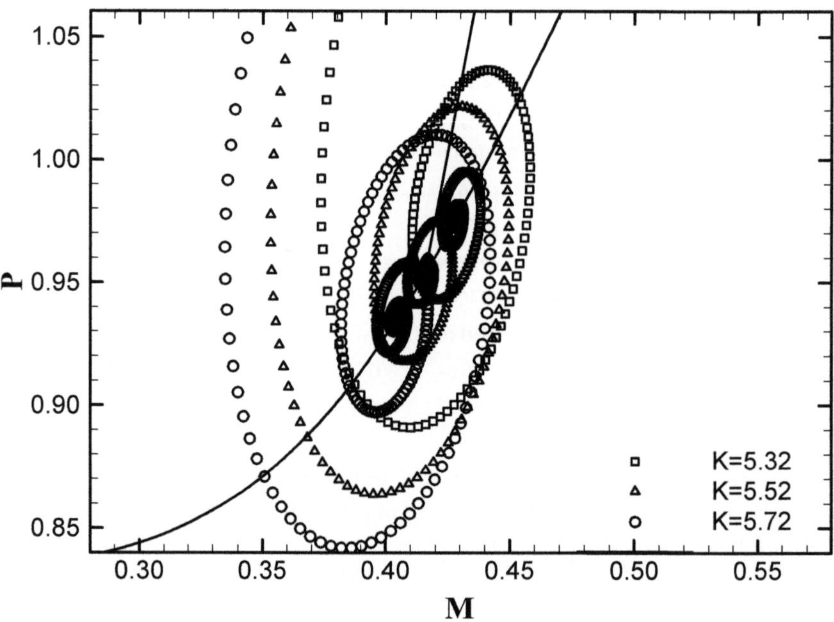

Fig. 5.2. Effect of changes of K on the cross point *(X, Y)*

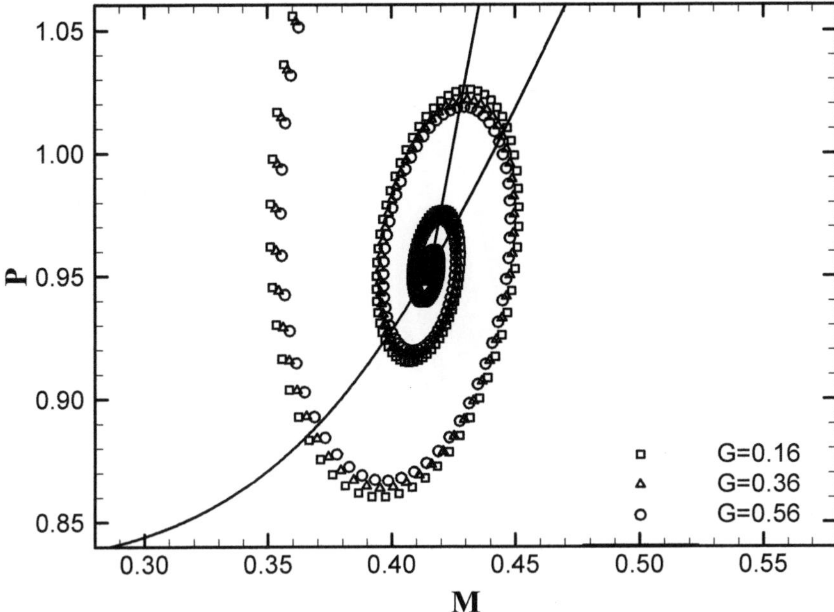

Fig. 5.3. Effect of changes of G on the cross point (X, Y)

5.4.1 Analysis for Case 1: with Parameters *B*, *G* and *K*

Referring to Table 5.3 for the results of the parameters B, G and K, it can be seen that parameters G and K affect the coordinates of the cross point (X, Y) most with parameter B having the least influence. It can be observed that the coordinates of the cross point (X, Y) shifts slightly upon the variation of values of K. As K increases, both the values of x–coordinate and y–coordinate of the cross point decrease (Figure 5.2). The values of the x–coordinate and y–coordinate will shift an overall of *2.25%* and *3.84%* respectively when K is changed, as indicated in Table 5.3.

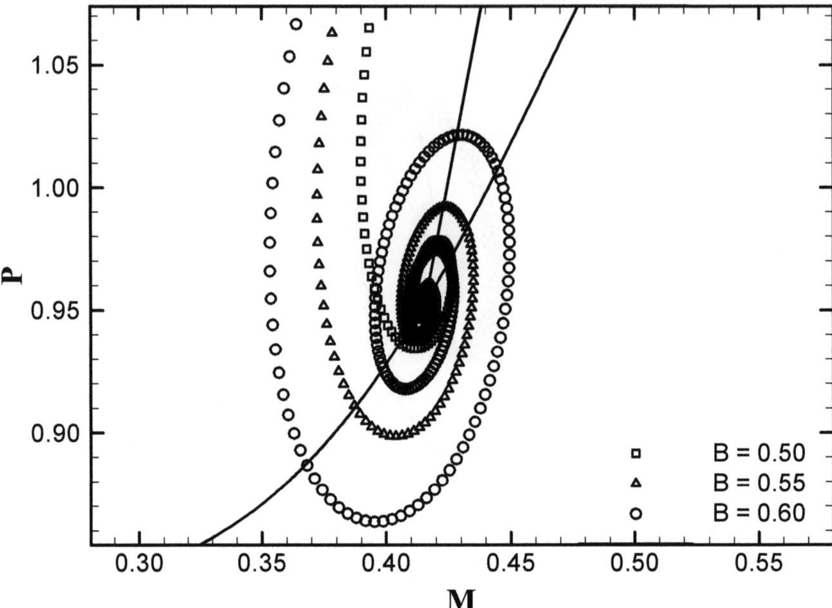

Fig. 5.4. Effect of changes of B on the cross point (X, Y)

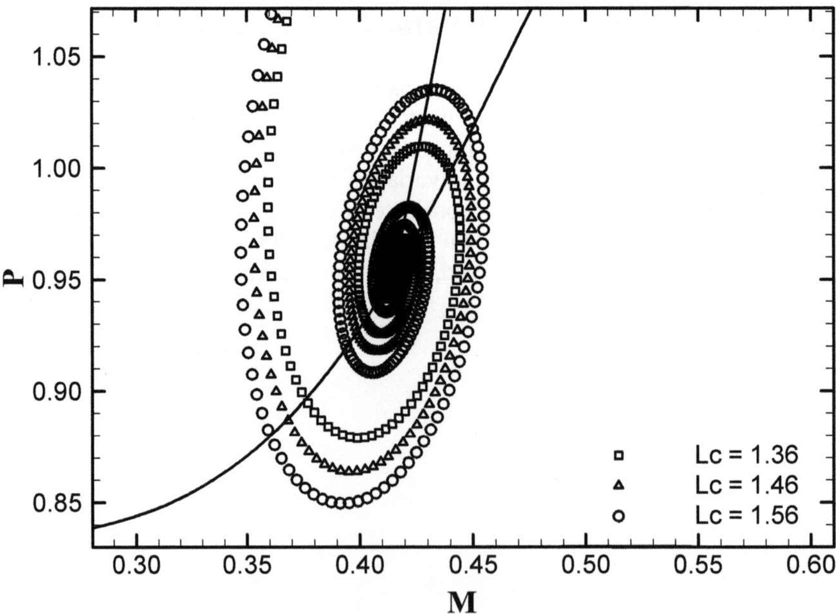

Fig. 5.5. Effect of changes of L_C on the cross point (X, Y)

On the other hand, from Table 5.3, the x–coordinate and y–coordinate will move an overall of *0.145%* and *0.262%* respectively when G changes. These small changes even cannot be observed in the figure (Figure 5.3). This implies that K has a greater effect on the position of cross point *(X, Y)*. The smaller influence of B can also be noticed from Fig. 5.4.

From Table 5.3, with a percentage of about *5%* and *10%* influence on the values of *Diff M* and *Diff P*, B is considered the deciding factor in this case. Hence it can be seen from Fig. 5.6 that upon increasing the value of B from *0.5* to *0.6*, the diameter of oscillation increases significantly from an average of *0.043* to *0.096* and *0.055* to *0.158* respectively for both the two axes. Comparing with B, K has a smaller effect on the diameter of oscillation.

On the other hand, looking at Fig. 5.7, if only G is increased with the rest of the parameters remain as constants, the increase in the diameter of oscillation is very small. When G increases from *0.16* to *0.56*, the average values for *Diff M* and *Diff P* for both the two axes decrease from *0.0745* to *0.062* and *0.115* to *0.093* respectively (Table 5.3). This results in an average value of *1.22%* and *2.11%*. Similarly, for parameter K, as shown in Table 5.3 and Fig. 5.8, there is a small average value of *1.55%* and *1.56%*. When compared with the percentage values for B, these values are much smaller.

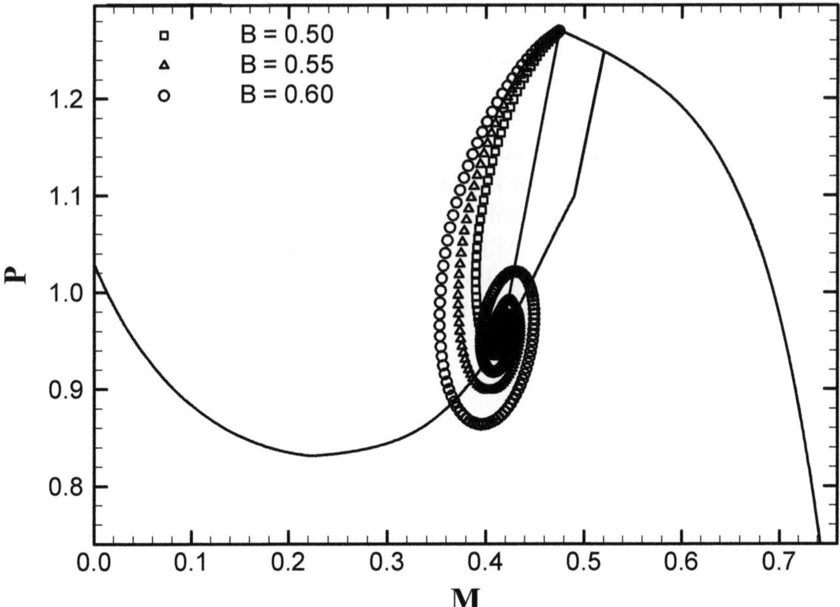

Fig. 5.6. Effect of changes of B on the diameter of oscillation

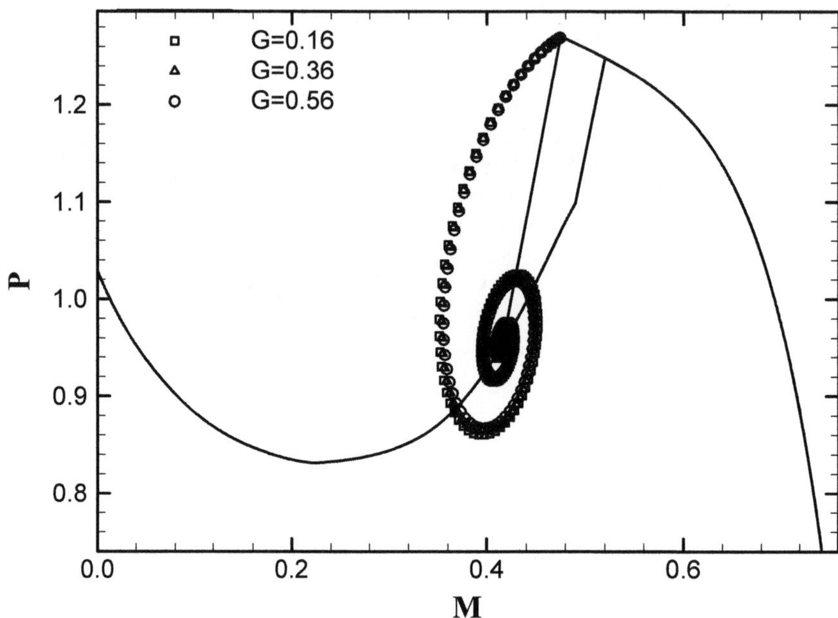

Fig. 5.7. Effect of changes of G on the diameter of oscillation

5.4.2 Analysis for Case 2: with Parameters B, G and LC

Similarly analysis is done for the second group of parameters B, G and L_C. Referring to Table 5.4, the last four columns complete the final values of the tabulation. As shown in the two columns for the percentage values, $R(X)$ and $R(Y)$, in affecting the coordinates of the cross point (X, Y), it can be deduced that parameters G and B are the two prime factors while L_C is the least important one. The locations of that the cross point (X, Y) when the values of G, B and L_C change are shown in Fig. 5.3, Fig. 5.4 and Fig. 5.5 respectively.

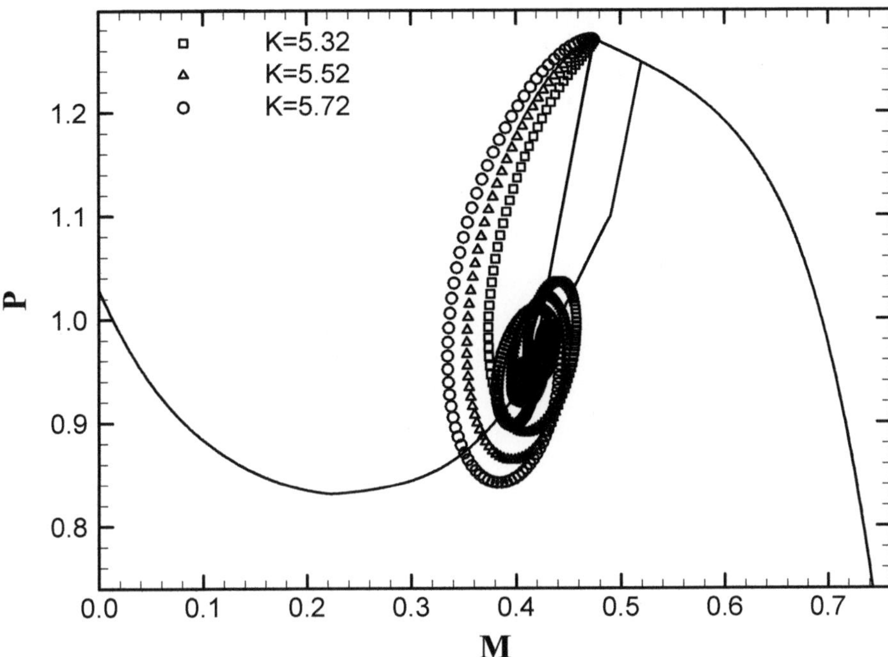

Fig. 5.8. Effect of changes of K on the diameter of oscillation

When B increases, the x–coordinate and y–coordinate of the cross point increases very little with changes in about *0.003%* and *0.035%*. Similarly for the parameter G when its value is increased, the percentage increase for both the coordinates is very small too, being *0.007%* and *0.023%* respectively. Though parameters B and G show some influence in affecting the position of the cross point *(X, Y)*, their effects are very small.

The last two columns of Table 5.4 indicate the percentage values of the diameter of oscillation for each parameter. It can be seen that parameter B has the highest value followed by parameter L_C and then parameter G.

The effects of parameters B and G on the diameter of oscillation have been analyzed in case 1 with Fig. 5.6 and Fig. 5.7.

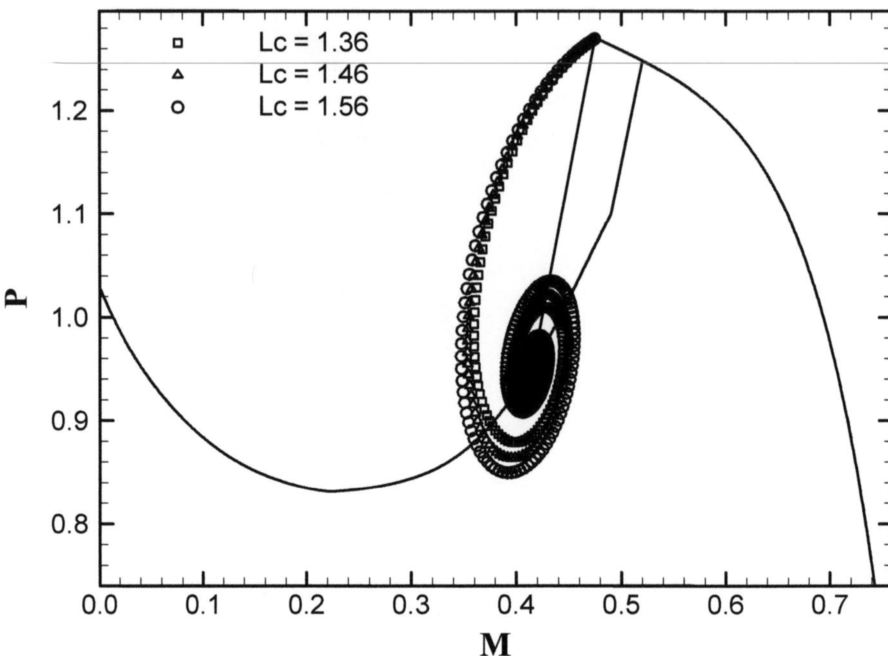

Fig. 5.9. Effect of changes of L_C on the diameter of oscillation

With reference to Fig. 5.9, the diameter of oscillation increases when increase the value of L_C. *Diff M* and *Diff P* increase from an average of *0.057* to *0.074* and *0.079* to *0.120*. This indicates an increase of *1.68%* and *4.08%* respectively. However, when compared with the percentage increase of *B*, those effects by L_C are still considered small. This implies that *B* parameter actually plays a more important role in the diameters of oscillations thus affecting the time taken for the system to reach a steady state.

5.4.3 Summary: Parameters B, G, K and LC

By mapping the results obtained from the two Taguchi Tables 5.3 and 5.4, a final analysis can be deduced.

It can be concluded that parameter *K* is the most important factor in deciding the position of cross point *(X, Y)*. It is then followed by *G, B* and L_C in the order of importance. From Table 5.3 and Table 5.4, the percentage values of changes in the *x*–coordinate and *y*–coordinate of the cross point of *K* is the highest with values of *2.25%* and *3.84%* thus implying that it affects the position of the cross point the most. L_C however, has the least influence with values at *0.0017%* and *0.023%* respectively.

Similarly, the analysis for the diameter of oscillation of the curve is obtainable. The most influencing factor is B followed by L_C, G and K. In conclusion, B has the highest percentage values and contributes a significant percentage in affecting the performance of the compressor system.

5.5 Conclusion

In this work, after reviewing the previous work on stall and surge of an axial compressor system and their mathematical models, the Greitzer's B-parameter model is applied to study the stall and surge characteristics. Thereafter, Taguchi method is used to further analyze the extent of the effects that the four parameters B, G, K and L_C have on the system. It is found that the Taguchi method succeeds in obtaining the important parameters that will affect the system.

The analysis on the four parameters, B, G, K and L_C has been carried out with satisfying results. It is observed that parameters B and K are crucial in maintaining the steady state operation of the whole compressor system. With much higher percentage value for the changes in $Diff\,M$ and $Diff\,P$, the parameter B highly affect the range of diameters of oscillation. On the other hand, K is the deciding factor in the position of the cross point with a higher percentage value in the variation of x–coordinate and y–coordinate.

In brief, analysis on the compressor system has been done by varying the values of B under the critical value (≈ 0.7). In Greitzer's model [5], analysis of the effect of B on the system has been carried out to varying values of B up to 5. However, as the Taguchi method has not been carried out previously on this system, analysis involving a greater range of B could be done as a further work. By varying the value of B from 0.7 (critical) onwards, the behavior of the system when it reaches surge can then be investigated.

References

[1]   Day, I.J., 1993, Stall inception in axial flow compressors. *Transactions of the ASME, Turbomachinery*, **115**: 1-9.

[2]   Day, I.J., 1993, Review of stall surge and active control in axial compressors. ISABE 93-7011, 97-105.

[3]   Day, I.J., 1994, Axial compressor performance during surge. *Journal of Propulsion and Power*, **10(3)**: 329-336.

[4]   Greitzer, E.M., 1980, Review–Axial compressor stall phenomena. *Transactions of the ASME, Journal of Fluids Engineering*, **102**: 134-143.

[5]   Greitzer, E.M., 1976, Surge and rotating stall in axial flow compressors. Part I: theoretical compression system model. Part II: experimental results and comparison with theory. *Transactions of the ASME, Journal of Engineering for Power*, **98**: 190-217.

[6]   Greitzer, E.M. and Moore, F.K., 1986, A theory of post-stall transients in axial compression systems, Part II: Applications. *Journal of Engineering for Gas Turbines and Power*, **108**: 231-240.

[7]   Jaw, L.C., et al., 1999, A high-response high-gain actuator for active flow control. *AIAA paper* 99-2128.

[8]   McDougall, N.M., Cumpsty, N.A. and Hynes, T.P., 1990, Stall inception in axial compressors. *Transactions of the ASME, Turbomachinery,* **112**: 116-125.

[9]   Moore, F.K. and Greitzer, E.M., 1986, A theory of post-stall transients in axial compression systems, Part I: Development of equations. *Journal of Engineering for Gas Turbines and Power,* **108**: 68-76.

[10]   Taguchi, G., 1993, Taguchi on robust technology development: Bringing quality engineering upstream. New York: *ASME Press.*

[11]   Taguchi, G., 1986, Introduction to quality engineering: designing quality into products and processes. *Japan: Asian Productivity Organization.*

[12]   Wilson, A.G. and Freeman, C., 1993, Stall inception and development in an axial flow aero-engine. ASME Paper 93-GT-2.

Appendix A. Programing: Greitzer's Model

This program solves the partial differential equation, (1), by using fourth-order Roung-Kutta method.

1.   The input data are

B   :   dimensionless parameter, $B = \dfrac{U}{2a}\sqrt{\dfrac{V_P}{A_C L_C}}$

G   :   dimensionless parameter, $G = \dfrac{L_T A_C}{L_C A_T}$

LC   : L_C, effective length of equivalent compressor duct

K   :   dimensionless parameter, $K = \dfrac{A_C^{\;2}}{A_T^{\;2}}$

MO   :   non-dimensional mass flow across compressor at stall limit
CO   :   initial non-compressor pressure rise
CD   :   non compressor pressure rise at stall
R   :   compressor rotor mean radius

2.   Initial condition:

The steady-state compressor pressure rise YC (or CSS) is obtained from the Cebyshev curve fitting
    CSS=CD, P=CO, C=CO, MC=MO, and MT=MO. Here CSS is the steady-state compressor pressure rise, P is compressor pressure rise, C is non-compressor pressure rise, MC is non-dimensional mass flow across compressor, and MT is non-dimensional mass flow across throttle.

3.   Output results can be found in the data files: OUT1.DAT and OUT2.DAT, in which, there are:

TS   :   the time scale, $\dfrac{TI}{2\pi}$, here TI is non-dimensional time

MC   :   non-dimensional mass flow across compressor
P   :   compressor pressure rise

Note:

- Please refer to the nomenclature about the explanations of variables in the code.
- Some variables are explained in the code after "!".

```fortran
      PROGRAM GRETZER_MODEL

*******************************************************************

*              COMPRESSOR GREITZER'S B-PARAMETER MODEL          *

*              METHOD: FOURTH-ORDER RUNGE-KUTTA                 *

*******************************************************************


      REAL CSS,T,TI,TS,CD,B,P,C,PI,H,LC,CO,MO,K,MC,MT,G
      REAL K1,K2,K3,K4,L1,L2,L3,L4,M1,M2,M3,M4,N1,N2,N3,N4
      OPEN (UNIT=1,FILE='OUT1.DAT')
      OPEN (UNIT=4,FILE='OUT2.DAT',STATUS='UNKNOWN')
C     SETTING THE INITIAL PARAMETER
      B=0.6
      G=0.36
      LC=1.46
      K=5.519


      MO=0.475
      CO=1.27
      CD=1.069


      R=0.2593
      PI=4.*ATAN(1.0)
      H=0.01
```

```
      NM=5027
C     INITIALIZE THE VARIABLE
      NR=2
      N=0
      TI=0
C     CALCULATE THE TIME LAD COEFFICIENT T
      T=PI*(NR*R)/(LC*B)
C     CALCULATE THE NUMBER OF SET OF OUTPUT DATA
      NF=NM+1
      CSS=CD
      P=CO
      C=CO
      MC=MO
      MT=MO
      WRITE(4,*) 'ZONE  I=',NF
      WRITE(4,6) MC,P
      WRITE(1,6) TI,MC,P
C     LOOP FOR ITERATION
201   N=N+1
C     RK 1ST IMTERMIDIATE CALCULATION
      K1=H*B*(C-P)
      L1=H*B/G*(P-K*MT**2)
      M1=H/B*(MC-MT)
      N1=H/T*(CSS-C)
C     RK 2ND IMTERMIDIATE CALCULATION
```

```fortran
      K2=H*B*(C+N1/2-P-M1/2)

      L2=H*B/G*(P+M1/2-K*(MT+L1/2)**2)

      M2=H/B*(MC+K1/2-MT-L1/2)

      N2=H/T*(CSS-C-N1/2)

C     RK 3RD IMTERMIDIATE CALCULATION

      K3=H*B*(C+N2/2-P-M2/2)

      L3=H*B/G*(P+M2/2-K*(MT+L2/2)**2)

      M3=H/B*(MC+K2/2-MT-L2/2)

      N3=H/T*(CSS-C-N2/2)

C     RK 4TH IMTERMIDIATE CALCULATION

      K4=H*B*(C+N3-P-M3)

      L4=H*B/G*(P+M3-K*(MT+L3)**2)

      M4=H/B*(MC+K3-MT-L3)

      N4=H/T*(CSS-C-N3)

C     RK FINAL VARIABLE CALCULATION

      MC=MC+(K1+2*K2+2*K3+K4)/6.

      MT=MT+(L1+2*L2+2*L3+L4)/6.

      P=P+(M1+2*M2+2*M3+M4)/6.

      C=C+(N1+2*N2+2*N3+N4)/6.

C     CALL FR SYSTEM DATA OF COMPRESSOR PERFORMANCE

      IF(MC.LE.0.475)THEN

C     CALL FOR COMPRESSOR PERFORMANCE

          CALL STALL(CSS,MC)

      ELSE

          CALL STY(CSS,MC)
```

```fortran
      END IF
C     UPDATE NON-DIMENSIONAL TIME TI
      TI=TI+H
C     CALCULATE TIME SCALE TI/2PI
      TS=TI/(2*PI)
C     UPDATE THE VARIABLE
      WRITE(1,6) TS,MC,P
      WRITE(4,6) MC,P
C     CHECK FOR EXIT ITERATION
      IF (N.LT.NM) GO TO 201
      WRITE (*,*)'CROSS POINT=',MC,P
6     FORMAT(1X,3F14.6)
      CLOSE(1)
      CLOSE(4)
      END


C     POST-STALL OF THE COMPRESSOR PERFORMANCE CURVE
      SUBROUTINE STALL(CSS,MC)
      REAL YC1,YC,XC,CSS,MC
      XC=MC
C     CALCULATE THE COMPRESSOR PRESURE CSS AT GIVEN FLOW RATE MC
      YC1=YC
      IF (XC.GT.0.222) THEN
      YC=1.775461-16.62889*XC+115.422724*XC**2-394.29835*XC**3
     &      +657.74784*XC**4-416.677009*XC**5
```

```
                              ! OBTAINED FROM THE CHEBYSHEV CURVE FITTING
          N=1
          ELSE
      YC=1.0285-2.31901*XC+11.6698*XC**2-36.0024*XC**3
     &          +64.5882*XC**4-37.9756*XC**5
                              ! OBTAINED FROM THE CHEBYSHEV CURVE FITTING
          IF (N.EQ.1) THEN
              YC=(YC1+YC)/2
              N=N+1
          END IF
      END IF
      CSS=YC
      RETURN
      END


C     PRE-STALL OF THE COMPRESSOR PERFORMANCE CURVE
      SUBROUTINE STY(CSS,MC)
      REAL YC,XC,CSS,MC
      XC=MC


C     CALCULATE THE COMPRESSOR PRESURE CSS AT GIVEN FLOW RATE MC
      YC=14.1545-120.092*XC+454.223*XC**2-868.284*XC**3
     &    +837.64*XC**4-327.125*XC**5
                              ! OBTAINED FROM THE CHEBYSHEV CURVE FITTING
```

```
CSS=YC

RETURN

END
```

Index

If you have any concerns about our products,
you can contact us on
ProductSafety@springernature.com

In case Publisher is established outside the EU,
the EU authorized representative is:
Springer Nature Customer Service Center GmbH
Europaplatz 3, 69115 Heidelberg, Germany

Printed by Libri Plureos GmbH
in Hamburg, Germany